HOW RADIO SIGNALS WORK

ALL THE BASICS PLUS WHERE TO FIND OUT MORE

JIM SINCLAIR

McGraw-Hill
New York San Francisco Washington, D.C. Auckland Bogotá
Caracas Lisbon London Madrid Mexico City Milan
Montreal New Delhi San Juan Singapore
Sydney Tokyo Toronto

McGraw-Hill

A Division of The McGraw-Hill Companies

1 2 3 4 5 6 7 8 9 0 DOC/DOC 9 0 3 2 1 0 9 8

ISBN 0-07-058058-8

This book was previously published by McGraw-Hill Book Company Australia Pty Limited.

Printed and bound by R. R. Donnelley & Sons Company.

This book is printed on recycled, acid-free paper containing a minimum of 50% recycled, de-inked fiber.

CONTENTS

PREFACE

This is a book for farmers and fishers, housekeepers and house builders, truck drivers and tax agents, sales people and anyone else who wishes to gain a basic understanding of the way radio signals work without being too much bogged down by technicalities.

It is presumed that you have no great knowledge of either electronics or mathematics and that you do not have any special interest in studying those subjects. It will be useful if you have an interest in the world around you to about the same level as an average layperson's understanding of the daily weather maps we see on television news broadcasts.

This is a pick-up-and-put-down book. You will find you can start reading at any one of a dozen places and find a fairly self-contained explanation of a particular aspect of the subject. The framework of the book is based on a method borrowed from computer programming where a short main program deals with the major topic and reference to subroutines is made for greater detail on particular aspects. There is also extensive cross-referencing so that once you start on a particular theme you can follow it through to its end.

OUTLINE PLAN

- Chapter 1 is an introduction to all other chapters.
- The section comprising Chapters 1 to 5 can be treated as a fairly self-contained introduction to the science of radiophysics.
- Any of Chapters 2, 3 and 4 can be treated as starting points if you have some outline understanding of the general subject.
- A full understanding of Chapter 5 requires previous knowledge of Chapter 3.
- In a superficial sense, each section of Chapter 6 can be regarded as a starting point with no prerequisites, but for full understanding you will need to be at least conversant with Chapters 2, 3 and 5.

- Understanding of Chapter 7 requires that you previously fully understand Chapters 2 and 3. Some knowledge of basic electrical theory may also help.
- Chapter 8 follows directly from Chapter 7 and has the same prerequisites.
- Chapter 9 requires that you are already fully conversant with Chapters 2 and 4.
- The section comprising Chapters 7, 8 and 9 is an outline of the 'nuts and bolts' level of technicality.
- Chapter 10 contains a number of independent sections which, in most cases, can each be treated as a starting point with no prerequisites.
- Chapters 11 and 12 can be treated at the superficial level as starting points with no prerequisites, but for full understanding you will require at least a general knowledge of all the matters raised in the previous ten chapters.

ABOUT THE AUTHOR

Jim Sinclair's fascination with radio began at age ten when he found a mouse-chewed World War II Air Force radio manual belonging to his father. By age sixteen he had gained his amateur radio licence. During the next few years Jim built his first transmitters and receivers and led an expedition to the top of St Marys Peak in the Flinders Ranges, South Australia, to test VHF propagation.

Jim became a Senior Technician on the installation crew of the first Radio Australia HF transmitter in Darwin. He spent six years on installation and maintenance of the HF and VHF radio telephone network at Alice Springs, covering a rugged and isolated five hundred kilometre radius. Extreme heat, floods and a helicopter rescue mission were part of his working environment.

Later in his career Jim carried out field strength and personal hazard measurements of broadcasting, television and radio communication facilities. He maintained broadcast transmitters and Channel ABAD7, Australia's first operational UHF television transmitter, and was Technical Installation Supervisor of the 100 kilowatt broadcast transmitter south of Alice Springs.

Jim has written this book to answer the many questions he is asked by friends and customers who wish to get maximum distance and reliability from their communications equipment.

Other publications by Jim Sinclair include *How Far Can You See* (a map reading template),* *User's Guide to Radio—Australian Inland Map*, *User's Guide to Radio—UHF Repeaters in Australia (Northern Territory)* and *User's Guide to Radio—UHF Repeaters in Australia (Victoria)*.†

---

* available from author (phone (08) 8383 6429).

† all user's guides are available from Graeme Electronics, PO Box 50, Mitcham, VIC 3132. Phone (03) 9873 4142, Fax (03) 9872 4229.

CHAPTER 1

THE RULES OF THE GAME

1.1 THE BASIC CONCEPTS

There are no prerequisites for this section.

In the list of subatomic particles, *electrons* are the smallest and lightest (and often the fastest) of those that are definitely known to be capable of independent existence. Electrons carry a negative electric charge, and when an electric current is detected it is usually the result of movement of a group or stream of electrons.

Most electrons spend most of their time attached to a particular atom. They move in orbit around the *nucleus* (which contains all the heavy particles). Fig. 1.1 shows a simplified diagram of a helium atom. There is more detail on atoms, electrons and the other subatomic particles in Chapter 10.

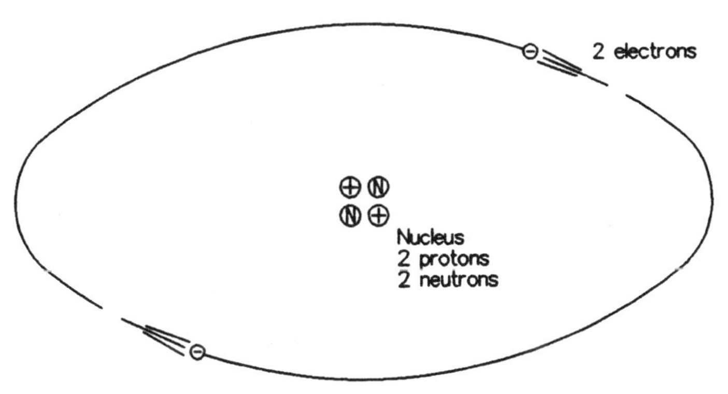

Fig. 1.1 *Helium atom (simplified diagram)*

Whenever an electron moves out of its orbit it has an effect on other electrons. Theoretically, one electron moving will have the potential to affect every other electron in the universe. In practice, all electrons are moving very often so the effect of one alone is usually detectable only by its nearest neighbours. Effects on electrons at greater distances may be gained by causing a number of electrons to move in unison.

The whole art of signalling by radio is to cause movement of a large-enough mob of electrons in unison in one place (the *transmitting aerial*) so that they have a detectable effect on electrons in another place (the *receiving aerial*).

The purpose of a radio signal is to carry information from one place to another—or many others in the case of a broadcast. If you seek to understand what is going on, you will need to understand half a dozen basic concepts which each by itself does not seem particularly relevant, but when you put them all together the subject suddenly falls into place. These concepts are:

1. what happens when magnetic/electric fields change;
2. vibrations, cycles and simple harmonic motion;
3. frequency, the spectrum of frequencies, and tuning;
4. modulation;
5. power transfer;
6. how signals of different types propagate.

The rest of this chapter carries a very brief outline of these concepts (along the lines of an extended summary). This is followed in the chapters immediately after by a more detailed treatment of each subject.

1.2 A BIT OF HISTORY

There are no prerequisites for this section.

Historically, radio and electronics have always been mixed together. The electric telegraph and wireless (radio signalling) were the mother and father of the electronics industry.

In the thermionic valves era, it was very difficult to control machinery using electronic circuits and the automatic control systems in use in those days were simple enough to be workable with mechanical relays. Up until the advent of the transistor, about the only things that electronics was used for were amplification of voices or radio transmission and reception.

The development of solid state (transistorised) analogues of various relay functions has sparked a massive expansion of electronics into all the modern-day fields of computing, digital data transmission, CAD/CAM and numerically controlled machining, and a host of other activities of that type.

A generation of bright young men and women has now grown up who have lived and worked with electronics all their lives and who know nothing about radio. The relatively few people who have attempted to study radio have found explanations of it buried in hundreds of pages of books that are basically about electronics. Most have decided it is all too hard and given it away.

Another more worrying trend that has grown in recent times is that some people who know little about the subject have taken some of the terminology and tried to equate it with spiritual, metaphysical or demonic forces.

Radio, that is, the transmission of energy and information through empty space, is not a hard subject to understand if you treat the electronics as black boxes with known functions. The properties of a radio signal can be completely defined by known and reliable physical forces which have no connection to the supernatural.

This book is about radio. It is not about electronics. It is not about the supernatural. When kept within those constraints the subject is not difficult to understand.

1.3 FIELDS—MAGNETIC AND ELECTRIC

There are no prerequisites for this section.

An electric current in wire gives rise to a *magnetic field*. This can be demonstrated by the effect of the current on a compass needle placed very close to the wire—the magnetic field exerts a physical force on the compass which tends to rotate the needle to a position perpendicular to the direction of the current (see Fig. 1.2).

An electrical voltage between two points gives rise to a field of electric force in the space between the two points. This *electric field* can be detected by the appropriate equipment almost as easily as can the magnetic field.

To demonstrate the principle we need only look as far as an experiment that almost all of us have done for fun as young children: rubbing a plastic pen or similar on the sleeve of a woollen jumper then using it to pick up tiny pieces of paper or dust and fluff (Fig. 1.3). This is the electric force in action.

These magnetic and electric fields exist because energy is temporarily transferred from the electrical circuit to the surrounding space. The fields of stored energy can be shown by theoretical calculations to reach an infinite distance in all

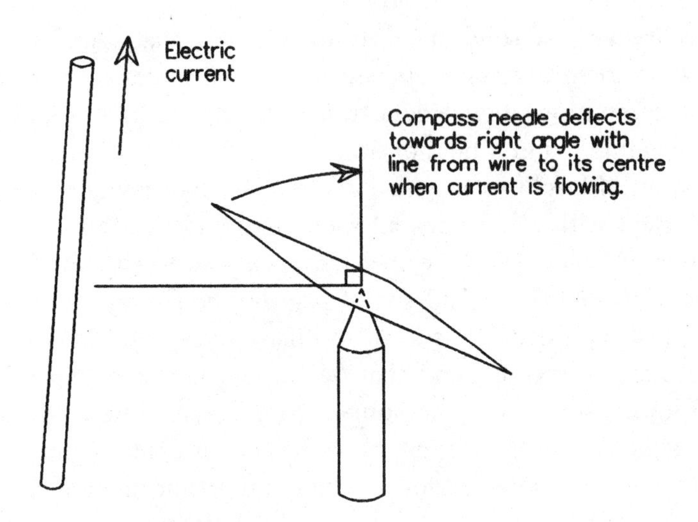

Fig. 1.2 *The magnetic effect of electric current*

directions—the concentration of energy per unit of volume is always decreasing with each move further away from the source but it is never quite zero. At the moment of commencement of the current or voltage, a pulse of energy radiates from the source to carry the energy stored in the field; when the current is stopped or the voltage removed, a similar pulse carries most of the energy back to the source. (There is more detail on this process in Chapter 3.)

No radiation of energy occurs while the electric current or voltage is steady; energy *is* radiated whenever changes are made to these. This can be demonstrated

Fig. 1.3 *The electrostatic force*

by placing a radio receiver tuned to a blank channel at a small distance from a wire which is carrying electric current and then switching the current. The steady current has no effect on the receiver but you should hear a click at each switching operation.

An experimental set-up in which two large plates of an electrically conductive material can be charged to high voltage and then the voltage switched on or off shows the same radiation of a click to a nearby receiver at each switch operation.

Continues Chapter 3: 'How energy is coupled'.

1.4 VIBRATION

This section is written presuming you have read Section 1.3.

To generate a continuous radio signal using electricity, the electric voltage or current must be continuously changing. To have a current continuously increasing is not practical. What *can* be done is to have the current increase for a short time then decrease back to zero, then increase and decrease, and for that cycle to be repeated as many times as is needed.

A continuous radio signal can also be generated by making a current flow into a piece of wire (an aerial or antenna) for a short time, then stopping it and making it flow back out again, and repeating that cycle for as long as is needed.

Continues Section 1.5.

1.5 CYCLES

This follows directly from Section 1.4.

All practical aerials are electrical conductors of a limited (finite) size. Some of them are rather big to look at but all are still finite in size. This means that any movement of electrons in one direction (electric current) will either fill or empty the conductor very quickly.

To radiate a continuous signal, the power from the transmitter must alternately push electrons into the aerial then pull them out again and continue this pushing and pulling for as long as the signal is needed. *One cycle* is one completed push and one completed pull.

For more detail you could go to:

Sine waves	Section 2.5
Resonance	Section 7.3
The driven element	Section 7.4

1.6 FREQUENCY

This section is written presuming you thoroughly understand Sections 1.4 and 1.5.
For each particular radio transmission there will be a certain number of cycles completed each second. This number of completed cycles is called the *frequency* of that transmission. It turns out that frequency is one of the most basic properties of a radio signal.

Tuning of a transmitter or receiver is achieved by changing frequencies. Frequency is the property that controls the pitch of a sound wave, the colour of a light wave, the hardness (penetrating power) of X-rays and gamma rays, and the band and channel of a radio signal.

It is only due to differences in frequencies that many signals can occupy the same piece of floor space at one time and be sorted out by the receiver. Differences of propagation are very largely dictated by the frequencies of different signals.

The unit of frequency is the *hertz*. A frequency of one hertz means that one cycle is completed each second.

hertz = cycles per second

Because electrons move so quickly, normal radio signals have frequencies in the range of many thousands or millions of hertz. The following abbreviations are used to make the numbers simpler:

kilohertz = thousands of cycles per second
megahertz = millions of cycles per second

1.7 THE RADIO SPECTRUM

This section is written presuming you have read Sections 1.4, 1.5 and 1.6.
Frequencies in common use in Australia range from 9 kilohertz up to about 22 000 megahertz. Signals in different parts of the range have vastly different characteristics.

In the same way that light of different frequencies has different colours and the whole range of colours is called the *spectrum* of light, the range of frequencies in use for radio signalling is called the *radio spectrum*.

Continues Chapter 2, 'Spectra'.

1.8 TUNING

To understand this section you will need to have read Section 1.6. Some basic knowledge of electrical principles would also be helpful.

There are electrical components and circuits which respond differently to signals of different frequency. In some cases the selective response to a small range of frequencies can increase the volume of a narrow band (about 1% either side of the reference frequency) by a factor of 10 to 50 times compared with all others in the spectrum.

A series of circuits, such as that arranged so that the circuits work on the signal one after the other as the signal flows through the transmitter or receiver, can be used to give whatever rejection of unwanted signals is needed so that the wanted signal is clearly heard or seen.

For more detail on frequencies, spectrum or tuning, read:

Wavelength	Section 2.2
Bands and channels	Section 2.4
Resonance	Section 7.3
Tuning and filters	Section 9.2

1.9 MODULATION

There are no prerequisites for this section.

The purpose of radio signalling is to carry some intelligent information. A steady stream of cycles of constant power and constant frequency is detectable as a radio transmission but does not carry any intelligence. A steady signal such as that would be described as an *unmodulated carrier wave*.

To carry intelligence, some property of that signal must be varied under the control of the program to be transmitted. *Modulation* is the name used to describe that variation under the control of a program. It turns out that there are only two properties of a radio wave that can be modulated:

1. The short-term flow of power can be controlled; this is called *amplitude modulation*.
2. The frequency of the wave can be varied over a small amount; this is called *frequency modulation*.

When the modulated signal is received, the intelligent message does not exist as a separate entity. The signal must be passed through a *detector* or *demodulator* circuit to reproduce the original message. There are many different detector circuits with minor variations in characteristics; however, the type of circuit must be appropriate for the type of modulation being received.

Continues Chapter 4, 'Modulation: the intelligent message'.
See also:
The diode detector — Section 9.5
Detecting other forms of modulation — Section 9.7

1.10 POWER TRANSMISSION

This section follows directly from Section 1.1.

Radio transmission is a power (or energy) transfer process. However, by normal power transmission standards, it is an extremely lossy process. A situation where one-million-millionth part of the power leaving the transmitter is coupled into the receiver would normally give an unusually strong clear signal. For a normal taxi or delivery van two-way radio system, a weak but workable signal would be about 10 000 times weaker than that.

Some of the weakest signals that have ever been detected are the radar reflections from the planet Venus. In that case, the signal returning to Earth was about another 1000 million times smaller again. This ratio is a number expressed by 10 followed by 25 noughts!

The ratio between power leaving the transmitter and power collected by the receiver is called the *path loss*. The important factor about it is that for any given pair of locations for transmit and receive aerials, the path loss is constant whether transmitter power is large or small so that variations due to modulation of the transmitter are faithfully reproduced in the receiver at the much smaller power level.

Because path loss numbers are so huge and small variations in them often do not matter a great deal, a logarithmic scale is used to express them. The *decibel scale*, originally devised for expressing gains and losses in the telephone system, is ideal for measuring path losses.

For more detail, read:

Path loss	Section 3.6
Why signals fade	Section 5.3
The decibel scale	Section 10.10

1.11 RADIATION IN FREE SPACE

This follows directly from Section 1.10.

Once the energy leaves the region of the transmitter aerial, it travels outwards at the speed of light. It is in fact radiant energy the same as light—only the frequency (and wavelength) is different.

The wave of energy may be generated purely by a change in a magnetic field or it may be entirely generated by a change in an electric field, but before it goes very far from the source it exists as a combination of magnetic and electric fields in a well-defined relationship.

In the region very close to the radiating element, there is a field of varying energy which can be easily defined as either a magnetic or electric wave. This is called the *induction field* and can be thought of roughly as energy being handed back and forth between the electric circuit and the field. Outside the induction field the wave loses any identity with its source, and irrespective of whether it started as an electric or magnetic field wave, it radiates as a combination of both.

To get past that bald statement and explain the nature of radiant energy is difficult because as with all the basic forces that drive our universe and hold it together, we humans can detect and measure them and to a certain extent predict their actions but we are totally and abysmally ignorant of their basic nature—all we know is that they happen!

When the ancients first observed electricity they named it literally 'the force of God' (*El* = God). After nearly 3000 years of research and experimentation we still have no more information than they did on the basic nature of these forces—why two objects not physically touching should be able to exert forces on each other. Their theory that these forces are direct evidence of the power of God in action is still as good as any other theory.

The radiated electromagnetic wave exists independently of the original current or voltage and will continue to radiate through free space (that is, any place where there are no conductors or absorbers) at the speed of light, even if the originating current is stopped and the conductor taken away. When the wave of radiant energy comes near a conductor, some of the energy will be absorbed by the conductor and set up currents and voltages that are miniature copies of those that started the radiated wave in the first place.

Continues Chapter 3: 'How energy is coupled'.
See also Chapter 5, 'How signals get there'.

1.12 A NOTE ON GRAPHS

There are no prerequisites for this section.
This book largely avoids mathematics; it is a 'What happens' rather than a 'How much' book. Radiophysics is, however, highly involved with mathematics so there must be some references to the subject. Wherever possible, those references to maths are presented as graphs.

Most people have at least a rough idea of what graphs are and how they work. The most widely used graphs are those which relate some activity to a time scale. There are some graphical presentations in this book for which the essential features are not obvious at first, although you may be able to identify them after some close study. For instance, there are some graphs later in the book where the x-axis is a single point!

For all graphs where the axes are two straight lines at right angles to each other, remember the following principles:

1. The x-axis is the straight line connecting all points for which the value of the y-factor is zero.
2. Each line which is parallel to the x-axis represents a particular value of the y-factor.
3. In almost all cases where the graph shows one factor being dependent on the other, the one that does the controlling is drawn along the horizontal x-axis. This is often called the *independent variable*.
4. The quantity that is dependent on (controlled by) the independent variable is drawn on the y-axis.

For example, in graphs of time-related factors, the clock just keeps ticking away without asking permission of anything or of anyone. It is the independent variable and so is drawn on the *x*-axis (or *x*-scale). The factor whose value depends on time is shown on the *y*-axis (or *y*-scale).

If you are not immediately sure of the meaning of a particular graph, check what factors are actually being displayed on the *x* and *y* axes. For example, all displays on the faces of cathode ray tubes (including television pictures) can be described as graphs and all are really a graphical comparison of two voltages. When an oscilloscope is being used to display an audio or similar waveform, one voltage is used to show the passage of time and the other shows the instantaneous level of the signal at each time, and the display is a graph of the waveform. By passing the original signal through a narrow band filter which can be varied in passband frequency, and using the frequency coding to control the voltage on one axis, the display becomes a spectrum analysis which is entirely different to look at but can be used to gain some of the same information about the signal as is shown by the first display.

The other item that can easily cause confusion about some of the graphs in this book is that in some cases the zero line is not always the line of 'no signal' or 'none of whatever is being measured'. Watch particularly in all cases where signal levels are being displayed using the decibel scale. Decibels are units of comparison and must always be referred to something else, and so the zero decibel line will usually be the line at which the signal being tested is equal to the reference.

SUMMARY

You will need to have at least a basic understanding of these points before any of the rest of the book will make much sense.

1. Energy is radiated whenever magnetic or electric fields are changed.
2. The radiated energy is in a form that disturbs the position of electrons in electrical conductors.
3. Energy radiation occurs only while the field is changing.
4. To generate a continuous radiated signal, the field must be continually changing.
5. A vibrating magnetic or electric field will generate a steady stream of waves of radiated energy.
6. Each vibration has a particular frequency and the waves passing any particular point will always show the frequency of the original vibration.

7. Different sources of radiation vibrate at different frequencies, and the resulting waves of radiated energy can be separated and identified by the frequency they carry.
8. To transmit an intelligent message, some property of the radiated wave must be modulated under the control of the intelligent program.
9. The short-term power level of the radiated wave may be modulated—this is called *amplitude modulation*.
10. The frequency of the wave may be slightly varied—this is called *frequency modulation*.
11. An electrical conductor placed in the stream of radiating waves will absorb energy from them which can deliver electrical power to the electronic circuit of a receiver.
12. Radio waves transfer power from the transmitting aerial to the receiving aerial but the proportion of power coupled is extremely small.
13. The coupling ratio for any given situation is constant so the signal in the receiving aerial is a faithful but much smaller copy of that transmitted.
14. Once the radiated wave leaves the transmitting aerial, it exists independently and continues to radiate even if the transmitting aerial is taken away.
15. The modulation of the radiated wave does not exist as a separate signal. The receiver must include a suitable detector to reveal the intelligent message.

CHAPTER 2

SPECTRA

2.1 THE ELECTROMAGNETIC SPECTRUM

This chapter follows directly from Section 1.7 and for full understanding requires that you have read Sections 1.3 to 1.7.

Radiation of energy using electromagnetic waves is known to occur with frequencies ranging from a few hertz (less than 50) up to millions of millions of megahertz. There is no theoretical reason why they should not occur outside that range but at present we cannot detect them.

Some electromagnetic waves affect our eyes and we know them as visible light. Others affect the outer layers of our skin in a way that we call either 'suntan' or 'sunburn' depending on the degree of the effect. Others we can feel as a warming effect from within our body. There are others that rip atoms apart which, if they strike us, will turn our body chemicals into poisons or change our genetics so that our children inherit mutations. There are others that will shine straight through our bodies but will affect a photographic plate and give us shadow pictures of our bones.

In all these cases it is the same radiated electromagnetic energy. Differences of the effect are entirely due to differences of frequency of the wave. Table 2.1 gives an outline of the effects of the different frequencies. It is drawn up assuming you understand Section 1.6, 'Frequency'.

The various forms of light and shorter wavelength radiations are usually specified by wavelength. Direct measurements of frequency are almost impossible, so if frequency of a particular radiation must be found, the wavelength is measured and the frequency inferred by calculation. The connection between frequency and wavelength is explained in Section 2.2.

Table 2.1 *Effects and uses of different frequencies*

Effects and uses of different frequencies
Frequency 25 to 400 hertz (Hz) Corresponding wavelength approx. 10 000–750 kilometres. Power mains frequencies. Some radiation is known to occur but is not used for any practical purpose.
Frequency 1 to 10 kilohertz (kHz) Wavelength 300 to 30 kilometres. Some pure research on 'VLF Whistlers'.
Frequency 9 kilohertz to 30 000 megahertz (MHz) Wavelength range 33 kilometres to 10 millimetres. Radio frequencies. Presently used for transmission of information of all types. The major subject of this book.
Frequencies above 1000 MHz are often named as *gigahertz*. 1000 megahertz = 1 gigahertz (abbreviated to GHz)
Frequency 30 to 3000 gigahertz (GHz) Wavelength range 10 mm to 0.1 mm. Radio frequencies not presently used to any great extent but potentially useful for short-range (within a large building) transmission of very high-rate data. Experiments are being conducted at present.
Frequency 3000 to approximately 430 000 gigahertz Wavelength range 0.1 mm to 0.7 μm (micrometres) (700 nanometres). Infrared light. Used for transmission of heat energy and some communication of very broadband data signals, mainly in optical fibre.
Wavelength 700 to 400 nanometres (nm) Visible light. Red corresponds to the 700 nm end of this range and blue/indigo/violet are at the 400 nm end of the range.
Wavelength 400 to 30 nanometres Ultraviolet light. The majority of photochemical effects (radiation interacting with molecules to cause chemical changes) are related to the visible and ultraviolet parts of the spectrum.
Wavelength less than 30 nanometres X-rays and gamma rays—the ionising radiations. In general, gamma rays are shorter than X-rays but there is some overlap. The major difference is that X-rays are those produced by electrical discharges in high-voltage equipment and which will affect a photographic plate; gamma rays are produced by radioactive decay and interact with the nuclei of atoms to change the number of protons in the nucleus, and this changes that atom to an atom of a different element.

Figure 2.1 summarises information about the electromagnetic spectrum. The rest of this book concentrates on that section of the spectrum from 9 kHz to 30 GHz which is the range that includes the radiofrequencies currently in use for transmission of information.

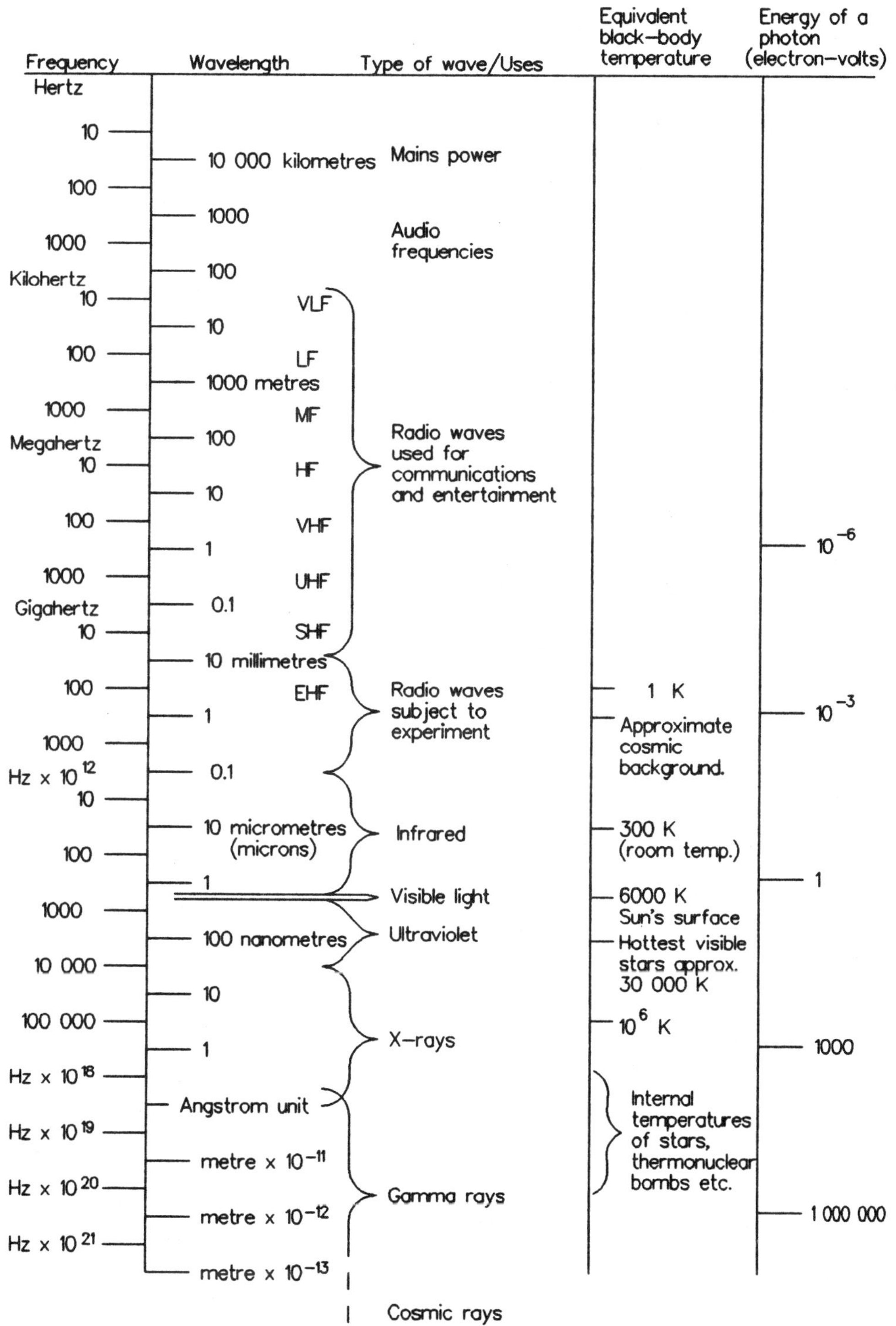

Fig. 2.1 *The electromagnetic spectrum*

2.2 WAVELENGTH

This section is written presuming you have recently read Section 2.1.

Light travels at about 300 million metres per second or 186 thousand miles per second. That fact is widely known. What is not quite so widely known is that all electromagnetic wave radiation travels at very close to the same speed. When a wave of any particular frequency is generated, the radiation from it travels at that speed; so for higher frequency signals, there are more waves in each second and so each wave is shorter than for low-frequency signals.

Most of us are familiar with the way bread can be cut either into thick slices or thin slices and we accept that there will be more of the thinner slices in each loaf (see Fig. 2.2).

Imagine a loaf of bread 300 000 kilometres long. (Imagination will do—please don't ever try to demonstrate this in practice!) If it is cut into 1 000 000 slices, each slice will be 300 metres thick; if it is cut into 300 000 000 slices each slice will be 1 metre thick.

In exactly the same way, a frequency of 1 000 000 Hertz (1 megahertz) will have a wavelength in free space of almost 300 metres; for a signal of 300 megahertz, the wavelength is 1 metre. Expressed mathematically:

Wavelength is inversely proportional to frequency.

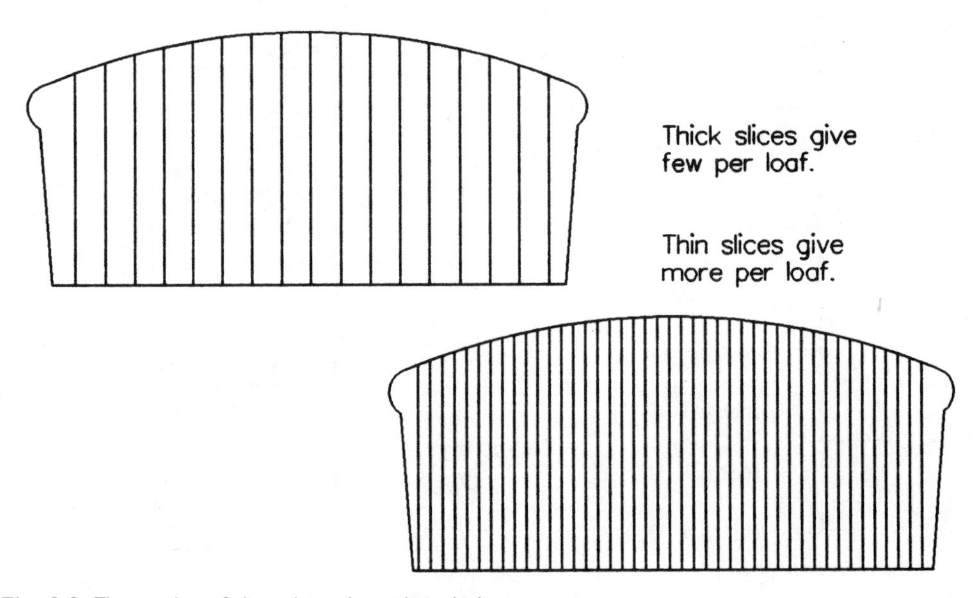

Fig. 2.2 *The number of slices depends on their thickness*

2.3 EFFECT OF MATTER ON WAVELENGTH

This follows directly from Section 2.2.
In this section the word 'matter' is used in its physics context. For an explanation, read Section 10.1, 'Matter, compounds and molecules'.

In empty space, radiation over all the spectrum travels at the same speed which, more exactly, is 299 700 kilometres per second. When matter gets in the way, the speed is slowed—only slightly for thin mediums such as air, but for more dense substances (such as water or some other transparent liquids) the speed may be only half to two-thirds of the free-space figure.

When radiation is slowed in a dense medium, the frequency stays constant and the wavelength is shorter for that part of the journey through the dense medium. This piece of information will be referred to again in Chapter 5, which deals with the propagation of radio signals.

When you are specifying a particular transmission by its position in the spectrum, the frequency is the factor that stays constant so it is frequency that gives the most exact measurement of a signal in relation to others.

2.4 BANDS AND CHANNELS

To understand this section you will need to have read these sections:

Frequency	Section 1.6
The electromagnetic spectrum	Section 2.1
Wavelength	Section 2.2

As mentioned earlier, the frequency of the carrier wave is one of the most important basic properties of a radio signal. It is possible to have carrier frequencies over a very wide range as shown in Table 2.2, and signals on different frequencies may have a wide range of different behaviours. To limit the range of possibilities that have to be considered at any one time, the part of the spectrum used for radio signalling has been divided into bands.

Note that the boundaries of these bands are entirely artificial. They were made up by consultations between tidy-minded bureaucrats and there is no particular physical process that marks the boundary between bands. The tidy-minded bureaucrats were, however, as well versed as possible in the radio technology of the day and they did arrive at a set of boundaries which are close enough to changes in physical conditions to be useful in everyday life.

Table 2.2 *The official bands*

Band	***Frequency range***
Very low frequency (VLF)	0 kHz–30 kHz
Low frequency (LF)	30 kHz–300 kHz
Medium frequency (MF)	300 kHz–3000 kHz (3 MHz)
High frequency (HF)	3 MHz–30 MHz
Very high frequency (VHF)	30 MHz–300 MHz
Ultra-high frequency (UHF)	300 MHz–3000 MHz (3 GHz)
Super-high frequency (SHF)	3 GHz–30 GHz
Extremely high frequency (EHF)	30 GHz–300 GHz

There are various other designations for bands with frequencies higher than 1 gigahertz. Capital letters such as 'S', 'L', 'X' or 'KU' are used to identify frequency ranges of interest for particular types of services. I have seen the term 'millimetric wave' used to describe frequencies between the SHF and infrared light regions, but it is not yet an official title.

Each of these bands represents a 10 to 1 frequency ratio which also means a 10 to 1 ratio of wavelength. Remember that the lowest frequencies correspond to the longest wavelengths. For example:

1. The low-frequency limit of the MF band is 300 kilohertz and this corresponds to a wavelength in free space of 1 kilometre.
2. The upper limit of the MF band (3 MHz) corresponds to a wavelength of 100 metres.
3. The UHF band ranges from 300 to 3000 megahertz which corresponds to wavelengths ranging from 1 metre down to 100 millimetres.

Each radio signal requires a definite width of spectrum within that range for its transmission. The name given to this quantity may be either *occupied bandwidth* (explained in Section 2.8) or *channel spacing*. Note at this stage that the occupied band of a particular signal may have little to do with the list of official band names shown in Table 2.2 earlier in this section.

The channel spacing required depends on the rate at which information is to be transmitted. For instance a Morse code signal at 25 words per minute may quite happily fit in a spectrum space of 100 hertz whereas a broadband bearer used for carrying a television program may require a bandwidth of 20 megahertz.

One common measure of channel spacing is the spectrum space required to fit one channel of normal voice transmission. The actual number of kilohertz needed depends on the type of modulation used and what the signal is intended to do, but where sections of the spectrum are set aside for voice communication, the channels are spaced for one voice to fit neatly in each channel.

For more detail see:

Sidebands	Section 2.7
Methods of modulation	Section 4.3
The 1 VF channel	Section 4.10

2.5 SINE WAVES

An understanding of Sections 1.3 to 1.6 is a prerequisite for this section.

For a number of practical reasons it is important that the carrier wave of each transmitter have one frequency and one frequency only. The shape or waveform of each individual cycle is important. It turns out that there is only one shape that can be transmitted as a single frequency. For most of the waveshapes you could imagine (such as squares, triangles, screw threads, the shape of sea waves), the energy of the wave can be analysed into a basic frequency, normally named the *fundamental*, and other frequencies called *harmonics* which are integer multiples of the basic frequency. (The integers are all the numbers without any fractions or decimal points. For instance 1, 2, 3, 4, 5 etc.)

The extra frequencies are real and not just mathematical fictions; a transmitter radiating one of these waveshapes would be heard by distant receivers on all of the calculated frequencies.

The waveshape that contains only one frequency is the shape that would be drawn by the device shown in Figure 2.3. To be strictly correct, the apparatus would need an extra mechanism like a clock escapement to supply power to make up for friction losses, but even without that, a well-made spring and a pen pressing lightly would show a curve close enough to the shape.

The single-frequency waveshape is mathematically described as the relationship between an angle and its sine. It could be drawn as a graph by drawing an x-axis along the middle of a long sheet of paper with angles from 0 to as far as you need to go along that x-axis. The y-axis covers all possible values of sine by ranging from $+1$ to -1.

The sine tables we study in high school show only one-quarter of the story: for angles from 0° to 90°. For angles from 90° to 180°, the curve moves back to the x-axis in a mirror image of the first quadrant. In the second half of the cycle from 180° to 360°, there is a curve exactly the same as the 0°–180° line but in the negative direction. This graph of $y = \sin x$ (see Fig. 2.4) is called a *sine wave*. It is the aim of every transmitter operator to transmit a carrier wave that is as near to a pure sine wave as possible.

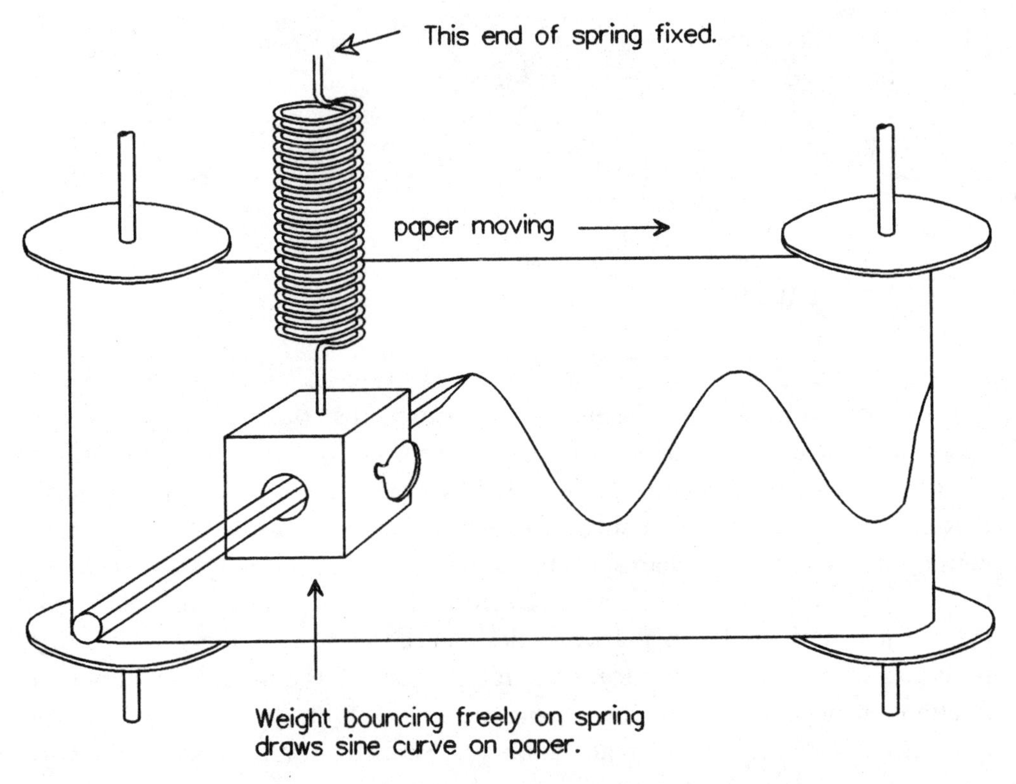

Fig. 2.3 *The device drawing a single-frequency waveshape called a sine curve*

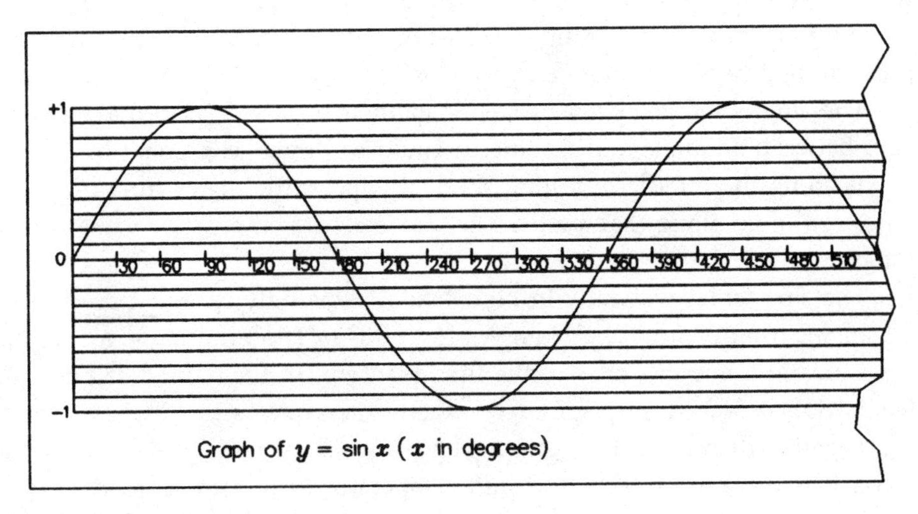

Fig. 2.4 *Graph of* $y = \sin x$

2.6 DEFINING 'CARRIER WAVE'

There are no prerequisites for this section.

From time to time in this book, the term 'carrier wave' is used, and it may be important that you have a clear understanding of exactly what this term means.

For a whole range of practical reasons, it is not possible to take an intelligent program, which could be audio or a television signal or a digital data bit stream, and simply amplify it to a suitable power level and send it to a workable aerial structure as a radio signal. To generate a workable radio signal, this intelligent program is modulated ('modulation' is described in Chapter 4) onto another signal which is a steady flow of cycles of constant frequency and power level. This second signal contains no intelligence of its own, but after the modulation process it carries the intelligence of the program. The non-intelligent signal of constant power and constant frequency is called the *carrier*.

A rough analogy can be made between a railway train and a radio signal in that the empty trucks can be equated with the carrier wave and the freight that the trucks carry represent the modulating program.

A closer analogy may be possible between the signal and the paper of a printer connected to a computer. The paper corresponds to the carrier wave and the printed text or graphics correspond to the modulating program. A string of line feed or form feed characters will produce a stream of white paper which corresponds to radiation of an 'unmodulated carrier wave' (see next paragraph).

You may be able to hear the carrier wave of a broadcast station during breaks in the program if your receiver is sensitive enough to give thermal noise when there is no signal at all. A signal that causes the noise to be silenced as you tune across it but does not produce a program is an *unmodulated carrier*. If your receiver has a *beat frequency oscillator* and you switch it on while you are receiving such a signal, you will hear a steady whistle or growl which, under good reception conditions, can be heard as a clear single-frequency musical tone.

2.7 SIDEBANDS

For full understanding of this section you will need to have read:

Cycles	Section 1.5
Frequency	Section 1.6
Sine waves	Section 2.5

Whenever two sine waves of exactly equal amplitude that are close in frequency but not harmonically related are combined, a pattern such as that shown in Figure 2.5 results.

Patterns of this type are common in modern life. You hear a *beat* when two musical instruments play slightly out of tune, when two similar motor cars move away from traffic lights, when a twin-engined aircraft is climbing after take-off before the pilot has synchronised the engines, and in many other places where two sounds are heard together. The 'dodge' tides seen on Adelaide's beaches and in a few other places in the world are caused by this effect.

There are some points worth noting about beat patterns of this type:

1. The frequency of the beat is the difference between the frequencies of the two waves.
2. The beat does not exist as a separate wave. If the two waves are electrical signals and you perform a spectrum analysis of the combination, you observe no power flow at the beat frequency.
3. If the two waves are exactly equal in amplitude, there is a point of exact cancellation at the trough of each beat. When the waves are close to but not exactly the same size, there is a deep trough with some residual signal in the bottom of the trough.

If three waves are combined under the right conditions, a similar pattern with a very significant variation results, as shown in Figure 2.6.

The conditions for producing this three-tone pattern are:

1. The tones are evenly spaced in frequency; for instance, 90 Hz, 100 Hz and 110 Hz; or 999 Hz, 1 MHz and 1.001 MHz. (Note 1 MHz equals 1000 kHz.)
2. The two outer tones are of equal amplitude.

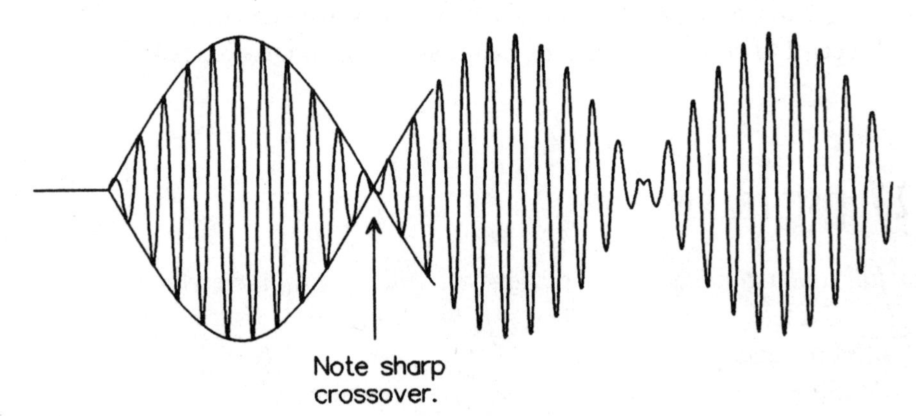

Fig. 2.5 *Two-tone pattern*

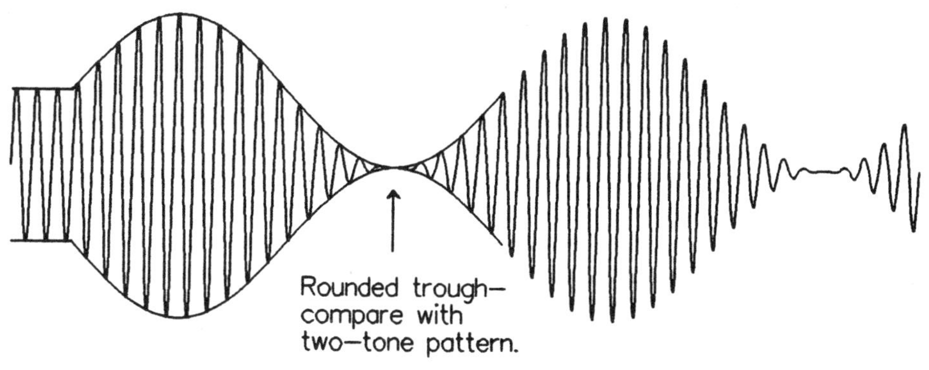

Fig. 2.6 *Three-tone pattern*

3. The central tone is exactly twice the amplitude of either of the others.
4. The two outer tones are related in phase so that their effects combine.

As in the two-tone example, the frequency of the beat is the difference, in this case, between the central tone and either one of the others. There is no power flow at the beat frequency; the beat does not exist as a separate signal.

The significance of the three-tone pattern is that it is exactly the shape of a radio frequency carrier wave 100% amplitude-modulated by a sine wave signal of much lower frequency.

If the three-tone pattern is taken to represent the voltage–time graph of an amplitude-modulated radio transmission, then:

1. the central frequency is the carrier wave;
2. the lowest frequency is called the *lower sideband*;
3. the highest frequency is called the *upper sideband.*

2.8 OCCUPIED BANDWIDTH

For a full understanding of this section you will need to have read all previous sections of Chapter 2 and, in addition, Section 1.6.

If a 1 kilowatt broadcasting transmitter (amplitude-modulated) was tuned to have a carrier frequency of exactly 1 megahertz and then set to transmit a steady tone of 1 kilohertz at 100% modulation, it would actually be radiating:

1. 250 watts of power on 999 kHz,
2. 1000 watts on 1000 kHz;
3. 250 watts on 1001 kHz.

These signals are not just mathematical fictions. If you tune a very narrow band receiver over that signal you can actually hear three separate signals which, in the case of a steady tone as above, are received as three separate unmodulated carriers. If the narrow band receiver is tuned slowly across a transmission of a speech program, it first gives clicking and pops in time with the sibilants of the speech, then a booming, distorted form of speech, then the carrier wave apparently with little or no modulation, then booming speech and finally the clicking of sibilants again.

In the process of amplitude modulation, extra signals are formed which have the effect of converting the steady carrier wave into the shape of the three-tone pattern as shown in Figure 2.6.

It turns out that every time a carrier wave is modulated, the modulation process generates sidebands. The range of sidebands may be different for different modulation processes but the general principle is that modulation converts a single-frequency carrier wave into a signal containing sidebands that occupy a definite width of spectrum space—a channel of a particular width.

The width of this band or channel depends on the highest frequency of the modulating program which in turn depends on the rate at which information must be transmitted. The bandwidth also depends partly on the method of modulation. Some examples are as follows:

1. The narrowest bandwidth in common use today is that of the CW (continuous wave) signals used to transmit Morse or telegraphic codes. Occupied bandwidth may be in the range 1 to 100 Hz.
2. The human ear hears sounds in the frequency range up to 16 000 Hz, so to transmit an amplitude-modulated signal with full fidelity, a bandwidth of 32 kHz is required.
3. The intelligence contained in speech is all concentrated in the frequency range from 300 Hz to about 2.5 to 3 kHz. To transmit that in single-sideband form requires a bandwidth of about 2.5 kHz or slightly less.
4. A television signal using the Australian PAL standard must transmit 625 lines of information 25 times per second, and ideally each line may contain up to 700 pixels (the smallest detail that can be transmitted) which could be alternately black and white. This requires almost 4 MHz which is increased to 5.5 MHz when sound and colour subcarriers are added.
5. The widest bandwidths in common use are found in the broadband data and multiplexed telephone links carried on point-to-point microwave bearers. Modulating frequencies up to about 20 MHz are currently in use.

2.9 BLACK-BODY TEMPERATURE

This section is not central to the explanation of signalling by radio but here is a piece of information that will be of interest to some readers. You may pass it over without losing the thread of the main story. This section has no prerequisites.

A physicist's definition of a 'black body' is a solid opaque object that is a perfect absorber or radiator of radiation of all frequencies. It is a fictitious object. All practical matter will give some reflection and there will be some frequencies which are not completely absorbed but some of the radiation will shine through. The theoretical fiction is, however, a useful reference point for calculations.

When a black body is heated, it is a source of radiation of a particular type. The characteristics of this *black-body radiation* are as follows:

1. There is no single frequency as with a radio wave. If the radiation is visible, we see it as white light and a spectroscopic analysis would show approximately even illumination over a range of colours. If the radiation is picked up by a receiver, the audio output is like the noise of rushing wind and no carrier frequency can be found. Signals of this type are technically described as *white noise*.
2. Examination of the radiation over a very broad range of the spectrum will show that there is a frequency at which the radiation is most intense.
3. For the spectrum on either side of the peak, the distribution of energy can be related to the mathematical Gaussian distribution of very large numbers of unrelated events.
4. The frequency of the radiation peak energy does not depend on any of the properties of the black body itself but is characteristic of the temperature and can be used to measure the temperature of the source of the radiation. This is the method astronomers use to measure the surface temperatures of stars.

For most real substances which are heated to incandescence, there is a thermal radiation which follows the black-body rules with modifications due to the chemical content of the substance.

The study and use of thermal black-body radiation is most relevant to visible light which corresponds to temperatures in the 2000 to 10 000 degree Kelvin range. We can for instance say, from observation of sunlight, and be quite definite about it, that if our Sun were at room temperature its colour would be black, and that its surface temperature is close to 5700 Kelvin. Now try telling someone that the Sun is black without explaining it in that rigorous detail and watch the funny looks you will get!

The measurement of temperature by the colour of the light is used mostly by astronomers and furnace operators (pottery, metallurgy etc.). For cooler uses, the 'colour' is in the infrared region of the spectrum—instruments exist for remote sensing down to the room temperature range but they are rare and expensive.

There is at least one measurement of temperature that is relevant to radio frequencies. The 'Big Bang' theory of creation suggests that there was a residual radiation, not connected with any matter source, which has the characteristics of a black body that is decreasing in temperature as the universe expands. At the present time this is thought to correspond to a temperature of 3 K. This 3 K cosmic background radiation corresponds to microwave radio frequencies and so broadband microwave detectors are being used to look for it. The search is the converse of Sherlock Holmes's classic story of the dog that did not bark. In this case the aim is to hear the bark then prove there is no dog. Hearing broadband microwave radiation is easy; proving that it is not from any physical source is much harder. This research will be continuing for some time yet.

CHAPTER 3

HOW ENERGY IS COUPLED

3.1 WHAT IS A MAGNETIC FIELD?

This follows directly from Section 1.3.

Most people are aware of the general principle that a magnet will attract a lump of iron and that if the lump of iron is small enough and the magnet close enough, the attraction will overcome gravity and the iron will be picked up. This force was first discovered several hundred years ago when it was noticed that if rocks of a particular type (lodestone) were hung on a very light piece of string, they would tend to turn around until they pointed in a particular direction (Fig. 3.1). That property was used by mariners who had a great need for a means of telling direction.

Compasses have been refined over the years. There are now highly accurate instruments available which can give directions to within a fraction of a degree and can even give a fair estimate of direction in a moving vehicle bouncing over a rough road or in a light aircraft being tossed around by thermals. Despite these refinements, all compasses rely on the fact that there is a magnetic field around the Earth which exerts physical force on magnetised materials.

As far as we can tell, the force is associated with the fact that electrons move in a spinning motion around the atomic nucleus. Magnetic forces are present whenever electrons move, but for most materials, the movements are disorganised so that forces in one direction are cancelled by equal and opposite effects in the other.

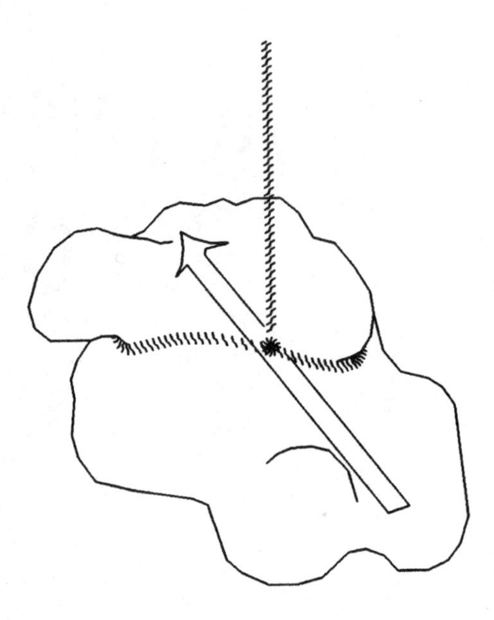

Fig. 3.1 *Lodestone*

In the few materials that show strong magnetic properties (iron, nickel and cobalt and some of their compounds), the spin directions of the electrons can be aligned so that there is a resultant force. In some materials such as soft iron, the electron spins can be *temporarily* aligned by an external magnetic field but will return to the disorganised state when the field is reduced. Other materials will magnetically align with an external field and, once aligned, will latch into the aligned state—these form the *permanent* magnets.

Energy is used to organise a magnetic field, and more energy is needed to change the state of the field of a permanent magnet. In the case of a piece of soft iron being magnetised by bringing it close to a bar magnet, the energy is supplied by the muscles of the person doing the moving. If the magnetism is generated by an electric current flowing in a coil of wire, the energy is drawn from the electric supply; it has the effect of slowing the rate at which the current builds up (see Fig. 3.2).

When the electric supply is switched off (when voltage drops back to zero on the graph), the magnetic field urges the current to keep flowing. The energy of the field is returned to the electric circuit and will cause a flow of current when no voltage is present.

Magnetism and magnetic induction can be a fascinating branch of physics to study and there is a lot more to it than this brief outline. However, it is a different subject from radio signalling and we must focus our attention onto one

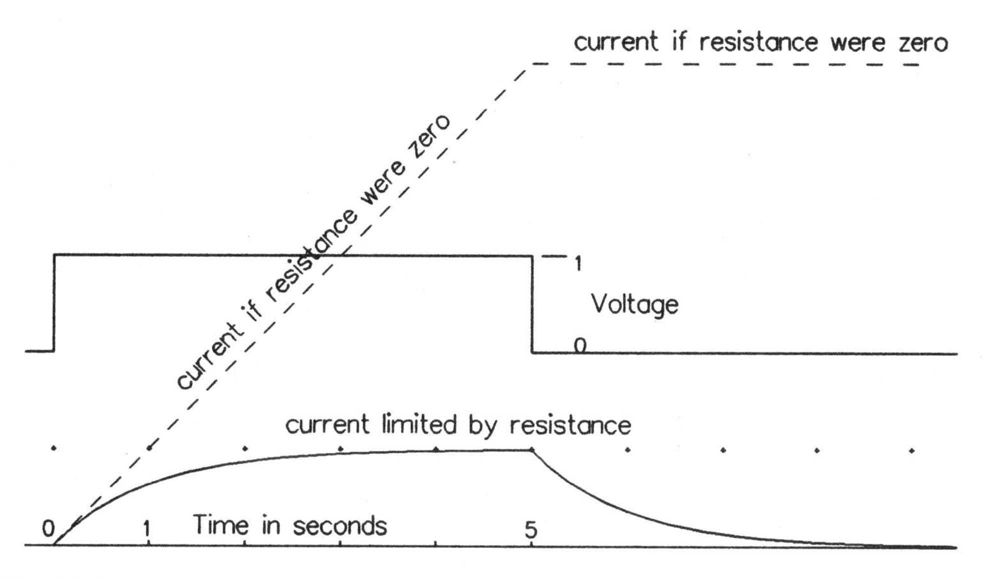

Fig. 3.2 *Switching graph*

aspect only: the mechanics of what happens when the field changes.

The magnetic field represents energy stored. In theory the field extends to infinity in all directions; in practice it becomes undetectable at a distance not much greater than the distance separating the two poles.

In cases where the magnetic field can be switched (for instance, one generated by an electric current), it is possible to show that there is a *wavefront* of energy moving out from the source due to switching on. When the magnetising current is steady, there is a static amount of energy stored in the field. When the magnetising current is reduced or switched off, energy is returned from the field. In this case there is also a wavefront which travels out from the source to signal the changed conditions.

You may be able to get some concept of this by thinking of water in a reservoir with a barrage or sluicegate, built into the wall of the reservoir, that can be opened very quickly. If the sluicegate is suddenly opened, a ripple travels upstream (Fig. 3.3).

The ripple is actually a 'wavefront' which has the shape of a step in the surface of the water. The wavefront is a starting signal for water to move downstream towards the new opening, but the wavefront itself moves upstream. In the magnetic field, the wavefront is a three-dimensional form of that effect.

When an electric current is steady, a magnetic field exists which contains stored energy, but there is no transfer of energy either into or out of the field.

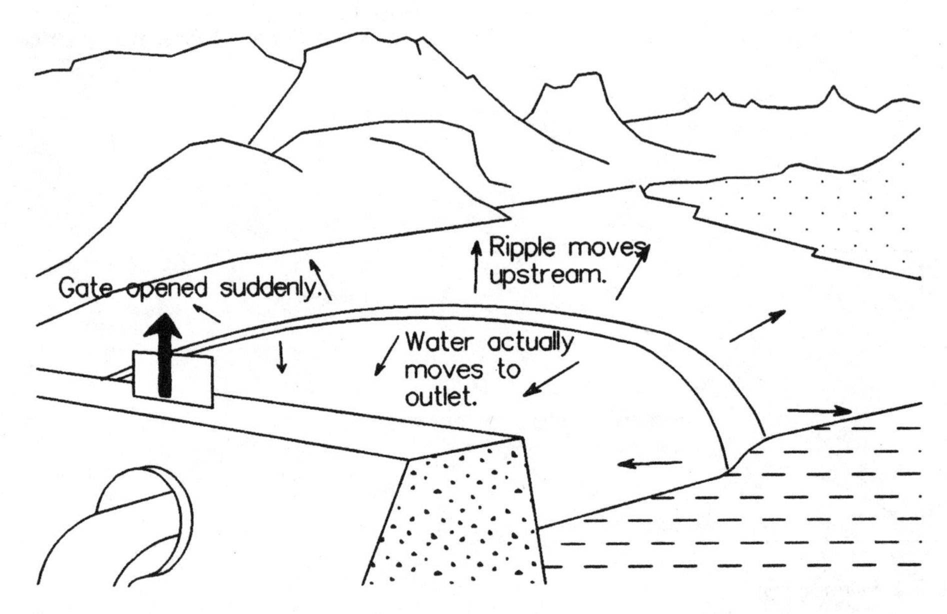

Fig. 3.3 *Wavefront ripple in a reservoir*

There are some materials which go into a state of *superconductivity* (that is, electrical resistance is absolutely zero) if they are cooled to within a few degrees of absolute zero temperature. If one of these materials is formed into a coil and held in a cryogenic chamber at cooler than the superconductive temperature, it will create a magnetic field when a power supply is connected and current is passed through it (Fig. 3.4).

If you short-circuit the ends of the coil in such an arrangement and take the power supply away, the current will continue to flow and the magnetic field will continue indefinitely. You have, in fact, a 'permanent' magnet for as long as the circuit stays connected and the temperature stays below the superconductive threshold.

3.2 WHAT IS AN ELECTRIC FIELD?

This section can be read with no prerequisites, but for a full understanding of it, you will need to read it in combination with Section 3.1.

If you suspend two light electrically conductive objects on fine wires close to each other but not touching, and then charge them so that there is a high voltage difference between them, you will find that there is a physical force attracting one to the other. For objects the size of kitchen saucepans and voltages such as

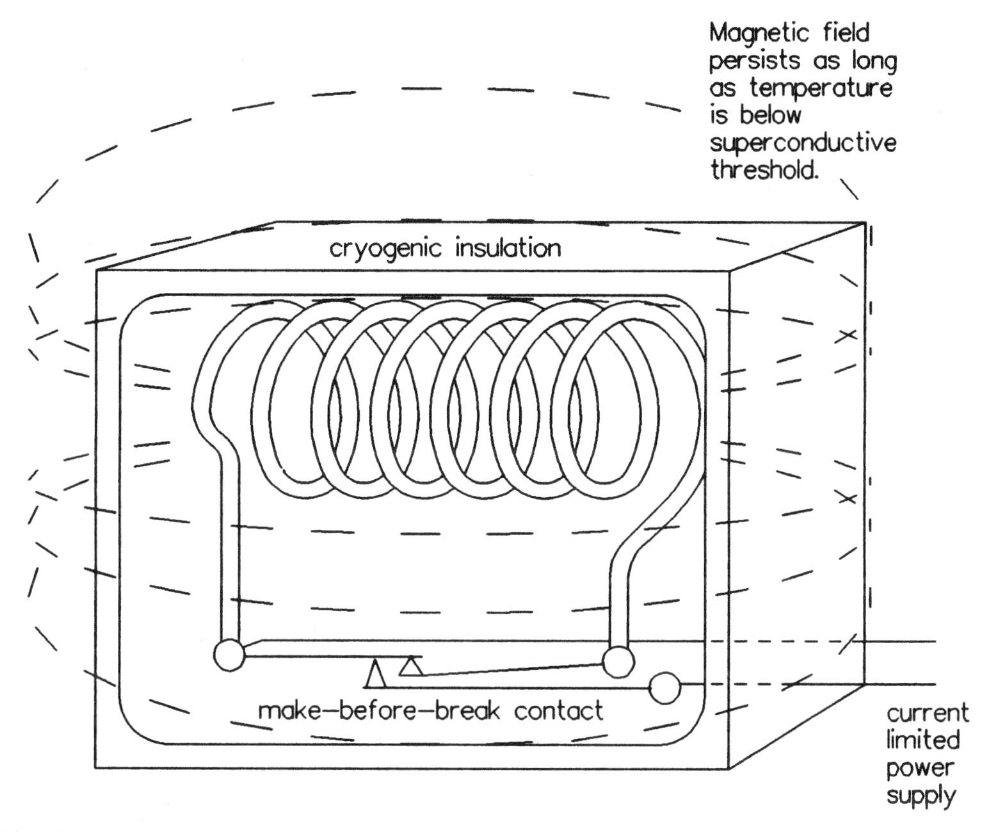

Fig. 3.4 *Coil 'magnet' in a cryogenic chamber*

those common in automotive systems, the force is minute and probably would normally be undetectable. If, however, the objects are made light enough, the suspending wire fine enough and the voltage high enough, the force can become comparable with the effect of gravity, and the change of position of the objects can be clearly seen with the naked eye.

This principle in the form of a device called an *electrometer* (Fig. 3.5) was the standard method by which early experimenters measured voltages. Today, mechanically refined versions of these devices are called *electrostatic voltmeters*.

There is a field of electrostatic force in the space between the charged objects and in the surrounding space; it has a lot of similarities with the magnetic fields described previously. The electric field also represents energy stored. Work must be done by the charging voltage (there is a flow of current for a short time while the voltage is building up) to create the field, and the voltage stays on the charged objects if they are disconnected from the power supply and left insulated.

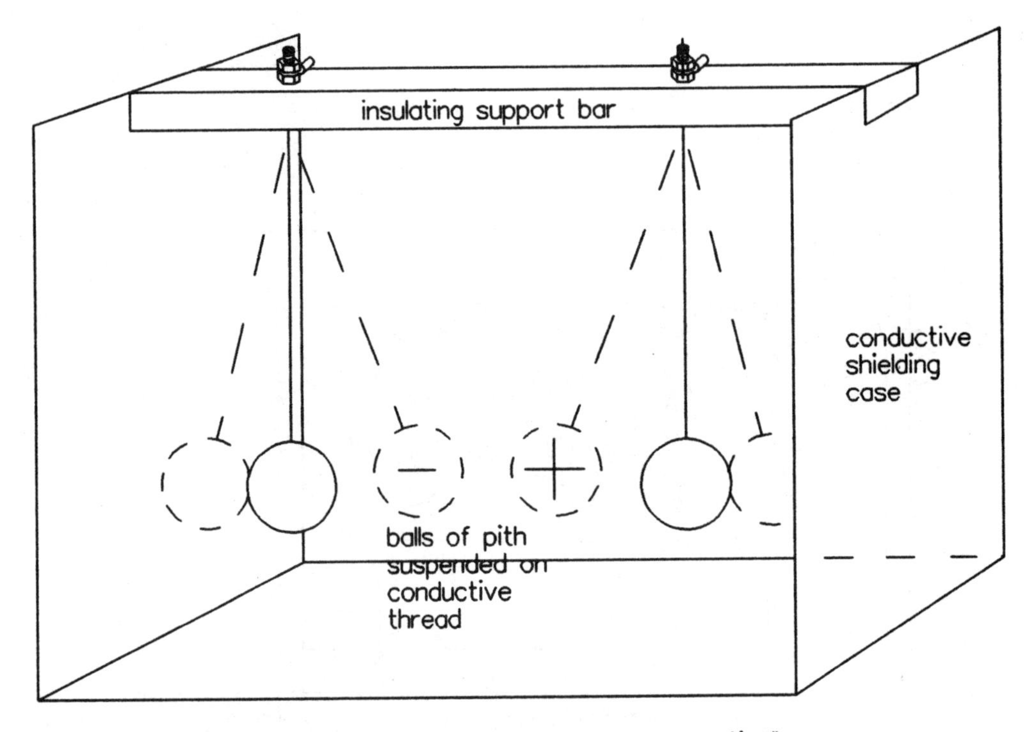

Fig. 3.5 *Electrometer*

Work is done by the field returning stored energy to the electrical circuit when the objects are discharged. This appears as a discharge current which flows for a short time after the connection of a discharging circuit.

As in the case of the magnetic field, there is an energy wavefront which travels out from the source to signal the commencement of the field, and when the source is discharged, there is a displacement wavefront which signals return of the stored energy to the source.

3.3 INDUCTION FIELDS AND RADIATION

To understand this section you will need to have read Sections 3.1 and 3.2.
In the next few pages the words 'energy', 'power' and 'force' are used frequently. In all cases they are used in the physical science sense. It may be important that you

understand what each of the terms means and, in particular, what the differences are between them. There are definitions in Chapter 10.

Either magnetism or electrostatics can be a fascinating basis for a life's work. There are scientific 'types' who spend all their life devising new experiments to test aspects of one or the other of magnetism or electrostatics, and whose only visible results are the publication of some scientific papers and possibly having their name used to describe one of the obscure but useful side effects. At the end of their working life, such people will say they have had a rich and rewarding time.

Both subjects are, however, different from radio communication. For the purpose of the present explanation we must focus attention on one aspect: what happens when the fields change.

In the volume of space immediately surrounding a magnet or an electrostatic charge, there is a field of stored energy which is capable of exerting a real physical force on an object (magnetic pole or insulated lump of electrical conductor) placed in the field. Each time the field changes, a wavefront travels out from the source to signal the change and adjust the energy storage of the field. If the field is continually changing (coil fed with alternating current or pair of plates fed with alternating voltage), there will be a continuous wave of magnetic or electrical power travelling out from the source of the field.

A suitable detector which is sensitive to the movement of magnetic or electrical energy rather than the steady-state field will show the wavefronts in the case of isolated changes and will give a steady reading in the case of alternating current or voltage. The question now is this: what happens when the detector is taken out of the field?

What we find is that even well away from the volume of space in which the steady-state field is detectable, there is an energy flow in response to changes in the field. The alternating field will give a steady flow of power which can be detected as alternations of the detected signal; a single switching will give a signal while the field is changing which may be heard as a click in a loudspeaker or seen as a short kick on a meter needle.

The region close to the source where steady-state magnetic or electrostatic forces can be detected is called the *induction field*. Outside the induction field where only the energy flow due to changes can be detected is called the *radiation field*. The boundary between the two fields is not well defined but there are some important differences in character between them. In the radiation field the energy flow is all outwards from the source, whereas the induction field contains some energy which will be returned to the circuit. An induction field may be either magnetic or electric with very little of the other type present; the radiation field is always a power flow of both magnetic and electric components in a well-defined relationship.

Because the magnetic and electric components are separate in the induction field, each one is subject, on its own, to the inverse square law (refer to Section 3.5), and the power flow which is the product of them tends to fall off in relation to the fourth power of distance from the conductor. In the radiation field, the inverse square law applies to the power flow. The instantaneous strength of the radiation field can be described mathematically as the 'differentiation' of the induction field. (See Fig. 3.6.)

Note that the electrical reality of 'differentiation' is a bit different from the purely mathematical relationship due to electrical resistance and the charge storage effect of capacitance (and there is always some capacitance in any circuit). The mathematical function specifies that one quantity depends on the rate of change of the other. As shown in the graphs of Figure 3.6 the output (resulting current) depends on the rate of change in the applied voltage in the strictly mathematical sense when the voltage/time graph is a slope, as shown on the right-hand side of Figure 3.6. However, when the change in voltage is a sudden step (which mathematically should result in an infinitely short pulse of an infinitely large current), the current actually rises instantly to a maximum value which is limited by resistance in the circuit and then continues for a short time after the voltage has steadied in a waveform of the 'half-life' shape which is typical of a discharging capacitor. The left-hand side of the 'resulting current'

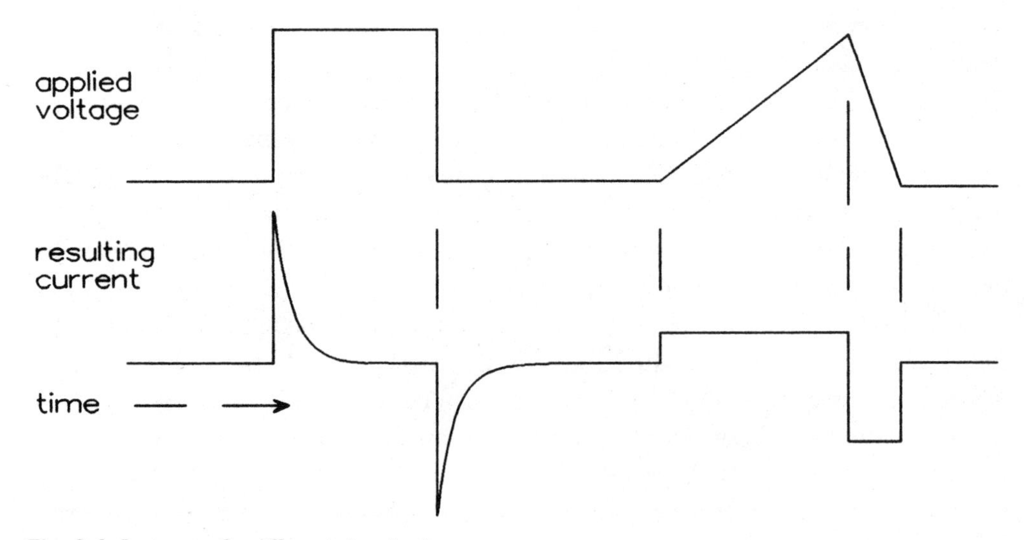

Fig. 3.6 *Response of a differentiating circuit*

graph of Figure 3.6 shows the way these stored charges affect the output of an electrical differentiating circuit. In theory a step change in the induction field should give a result that is mathematically exact but in practice the only instruments we have to measure the effect are electrically based, so all practical measurements will show waveforms similar to that of Figure 3.6.

3.4 MEASURING POWER FLOW—THE ISOTROPIC RADIATOR

This section follows almost directly from Section 3.3.

Consider the case where the source circuit has a continuously varying (alternating with a sine wave) current or voltage. In the radiation field this will be detected as a continuous signal and can be made to give a steady reading of power flow on a suitable meter. The question is: How much power is flowing?

There is energy in the induction field which, in the case of an alternating current or voltage, is swapping back and forth between the circuit and the field so that for any time longer than a couple of cycles of alternation, the net power flow will be close to zero. There will be some losses of power in the electrical circuit itself due to resistive losses. In a circuit designed for efficient radiation, this will be a small part of the total electrical power (volts × amps) flowing. All of this loss appears as heat, so for critical measurements, the loss can be calculated from the temperature rise of the components. The majority of the electrical power flowing into the radiating circuit appears in the field as a radiated signal.

Imagine now that a source of radiation could be placed at the exact centre of a large hollow sphere and that it is able to give exactly even radiation to all parts of the inner surface of the sphere. In practice such a situation is not possible; like the black-body radiator, this is another mathematical fiction that provides an easy basis for comparison of more complicated practical structures. This one is called an *isotropic radiator*.

The value of the isotropic radiator is that we know what power is being radiated, and if we can be sure that it is evenly spread around the sphere, we can by simple division calculate the power density per unit small area of the sphere's surface. Practical aerial structures do not give even illumination of the whole sphere, but for simple arrangements, the difference from isotropic can be calculated and this is a useful reference for comparison. When the power flow does not give exactly even illumination of the whole sphere, we can still say that the total of the power radiated is what is falling somewhere on the sphere.

If the surface of the sphere is divided into many small sections and a measurement made of the power in each section, the total of all sections combined must equal the power supplied by the radiator being tested. This fact can be used in the laboratory as a check on the calibration of both measuring instruments and test signal sources. When practical aerials are being tested we can say for sure that if intensity of signal is reduced in some directions, then in the directions where signal is maximum it will be stronger than it would be if the radiator were isotropic.

For more information see:
The driven element Section 7.4
Antenna gain and directivity Section 7.7.

3.5 RECEIVING AERIALS—CAPTURE AREA

To understand this section you will need to be familiar with the principles defined in Sections 3.1–3.3 and also have some understanding of the concepts of 'frequency' and 'wavelength' as defined in Chapter 2.

The concept of the radio signal illuminating the inside of a hollow sphere, as described in Section 3.4, is useful for working out what happens to signals travelling where there is nothing in the way. The same 'inverse square law' as is used for visible light measurements applies equally well to radio signals and for the same reason: the power is shared over a surface area which depends on the square of the distance from the source (Fig. 3.7).

When an aerial structure is used for receiving, it behaves as if it were collecting all of the power falling on a particular area and transforming it into electrical energy available at the output terminals. In a test situation, if the illumination in power per unit area (of the hollow sphere) can be measured or calculated, then a measurement of the maximum power that can be drawn from

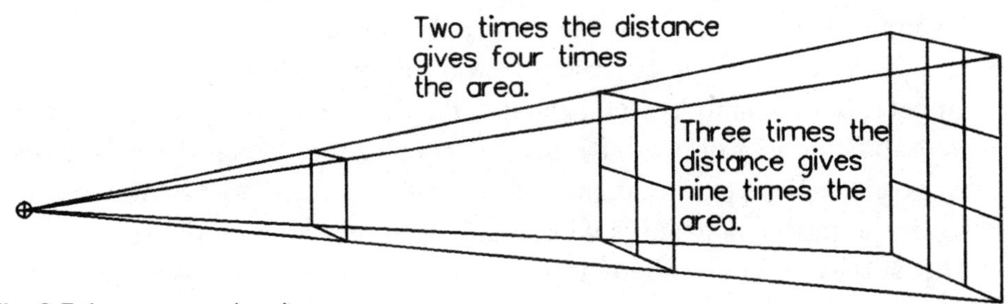

Fig. 3.7 *Inverse square law diagram*

the output terminals will indicate that the structure under test is equivalent in effect to a perfect absorber of a particular surface area. *Capture area* is the name given to that equivalent surface area.

For high-gain structures such as satellite dishes where the physical size is many wavelengths across, the capture area is closely related to the physical area and in many cases can be closely estimated with a tape measure. On the other hand, for single-element wire antennas, the presence of the conductor has an effect on field conditions for some distance around the wire, so physical surface area is not at all related to capture area.

For simple receiving structures, capture area may partly depend on wavelength to the extent that the length of the element must be matched to wavelength for tuning; however, it is not possible simply to measure the physical dimensions and calculate capture area without a lot of detailed knowledge of the operation of the device.

Measurements of capture area are most useful in the cases of parabolic dishes and horn antennas (Fig. 3.8), where the aim is simply to intercept the power of a certain area of the incoming signal and guide that power to the input terminals of a box of electronic circuitry.

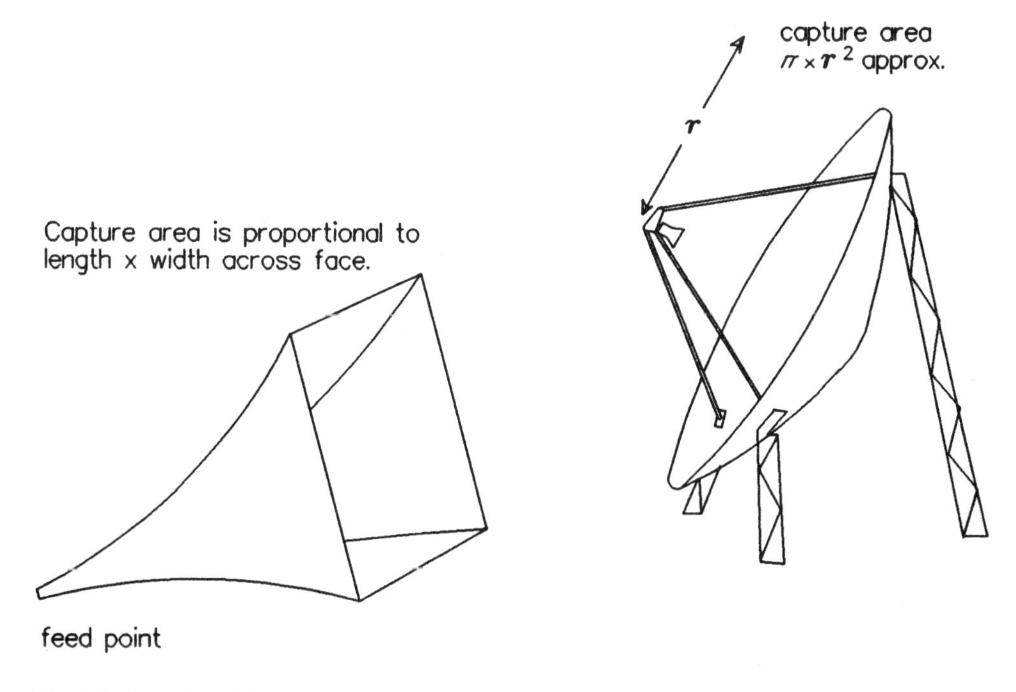

Fig. 3.8 *Examples of high-gain antennas*

3.6 PATH LOSS

The only prerequisites for this section are Sections 1.3 to 1.5. However, for full understanding, you may also need to refer to points in any of the early sections of this chapter.

The purpose of a radio system is to transfer energy from the electrical circuit of the transmitter to the electrical circuit of the receiver then to use that transferred energy to carry an intelligent signal. In the process of radiating energy and intercepting the radiation with a receiving aerial, it is accepted that most of the radiated energy will be lost and only a minute proportion will be received.

In most cases, when the receiver is moved further away from the transmitter, the dispersion of radiated energy is made greater and the proportion received is smaller. The limitation on maximum range of a practical radio system, then, is related to how much loss can be tolerated before the signal becomes unworkable. The maximum range depends on:

1. electrical power output of the transmitter;
2. maximum sensitivity of the receiver;
3. efficiency and directional properties of the radiators;
4. propagation conditions for the signal.

Each of these factors will be considered in more detail in later chapters.

If there were nothing in the way (i.e. outer-space vacuum) and nothing close to the direct line between transmitter and receiver, it would be possible to make a simple calculation based on the hollow sphere and capture area principle (see Sections 3.4 and 3.5) which would give a figure for the strongest signal that could ever be expected in a particular situation. The loss that applies between the two radiators is called *free space path loss*. This calculation is often done as a first step in more detailed calculations on performance of point-to-point links.

Another simple calculation comparing transmitter output with the weakest usable signal at the receiver gives a figure for *maximum allowable path loss*.

If the proposed link is such that free space path loss is greater than the maximum allowable, the system has no hope of working under any conditions. The amount by which free space loss is less than maximum allowable loss gives an indication of how much can be allowed for the combined effect of propagation losses and margins for fading.

The detailed engineering of the link then calculates expected propagation losses. What is left over is a margin for fading. The final decision on whether it is workable in that form is a statistical calculation to translate that fading allowance into a proportion of time during which the link will not be available. Calculations of this type may be described as *link budgets*.

If the calculations show an unworkable condition, the designer can specify either higher transmitter power, more directional aerial structures, or higher towers, and then repeat the calculations. Occasionally there may be an option to specify higher sensitivity receivers, but the initial specification will usually be for the best that is available; there is hardly ever a need to purposely use a low-sensitivity receiver.

To put some typical figures on these principles, the following could be taken as expected ranges of path loss for voice services operating in the VHF and low-end UHF range of frequencies over distances up to 40 km:

1. Free space path loss between aerials is often in the range of 80 to 120 dB (decibels).
2. Maximum allowable loss from transmitter output terminal to receiver input terminal is usually about 140 to 150 dB. This may be increased to 160 to 180 dB when directional aerials are added.
3. The margin for fading required depends on the order of reliability required, varying from about 25 dB for 90% reliability up to perhaps 40 dB for 99.9% reliability.
4. Thus the amount available for losses over obstacles may vary from almost nothing in cases where low-power transmitters are required to give highly reliable service, up to perhaps 40 to 50 dB in cases where higher power is available and lower reliability can be tolerated.

Path loss will be mentioned again in Chapter 5, 'How signals get there'.

3.7 EXPRESSING FIELD STRENGTH MEASUREMENTS

This section extends the detail of Sections 3.5 and 3.6.

Thinking of a radio signal as a certain amount of power flowing through a particular area in free space is a convenient way of defining the conditions that apply to a fixed link using microwave frequencies and parabolic reflector or horn radiator arrays. In other types of service, however, an attempt to calculate a signal strength in terms of power flow gives an unwieldy calculation with little relevance to practical situations.

Taking the case of a general coverage receiver operating on a range of frequencies below 30 MHz using a random length wire as an aerial, the figure of interest is the number of microvolts of signal that can be presented to the aerial terminal of the receiver. In these cases it is much more convenient to express the field strength as microvolts or millivolts per metre which is a figure that gives a direct relationship between the length of wire and the signal strength at the receiver input.

Note at this point that a field strength quoted as 1 millivolt per metre does not mean that a 1 metre length of wire will deliver 1 millivolt to a receiver; the figure defines the source electromotive force and would only appear if the receiver presented no load at all to the aerial, that is, if it had infinitely high input impedance. In all practical cases, the receiver input does present a load and for electrically short pieces of wire (less than a quarter-wavelength), there is a high source impedance which reduces the voltage of the available signal.

In all cases, field strength expressed as microvolts per metre can be related to field strength expressed as microwatts or picowatts per square metre, but it is a complex calculation involving the electric and magnetic field vectors and the relationship between them. If you have a need to know the details of this calculation, a good source of the relevant formulae would be one of the more comprehensive books on engineering principles which are held in libraries under classification number 621.384.

SUMMARY

1. Steady magnetic or electric fields contain stored energy but none is being radiated while the field is constant.
2. Changes to the field cause energy radiation. A field that is constantly changing, such as a magnetic field resulting from an alternating electric current, gives continuous radiation of energy.
3. For the purposes of calculation, a mathematical fiction called the *isotropic radiator* is used to compare the directional properties of aerials and antennas.
4. Preventing radiation in some directions will cause a more concentrated radiation in others. This fact can be used to increase signal strength at a distant receiver.
5. If there are no obstacles in the way, the radiated energy obeys the inverse square law of optics.
6. At the distant receiver, the radiated energy will induce electrical voltage into conductive elements, and currents will flow in response to those voltages.
7. The electrical current can be led away to an electronic circuit capable of amplifying and detecting the radio signal.
8. The power delivered to the electronic circuit depends on the power output of the transmitter, the arrangement of conductive elements in the transmitting and receiving aerials, and the path loss between them.

9. The receiving aerial behaves as if it were a perfect absorber of energy of a particular surface area perpendicular to the direction of radiation.
10. The proportion of power intercepted by the receiver is extremely small. A proportion of 1 part in 10 000 000 000 000 000 (10^{16}) would give a typical signal strength in many situations.
11. For any given situation, the proportion is constant so that a change in power at the transmitter results in a corresponding change at the receiver.

CHAPTER 4

MODULATION: THE INTELLIGENT MESSAGE

This chapter is an expansion of the introductory Section 1.9 which, in effect, is a definition of the term 'modulation'. You should read Section 1.9 before you get too deeply involved in the detail of this chapter.

4.1 WHAT IS A VOICE?

There are no prerequisites for this section.

When we speak, we vibrate our vocal cords in particular ways so that they will transmit the vibration to the surrounding air particles. The vibration of the air particles then radiates from our mouth as waves of compression of the air (Fig. 4.1). If there is an ear within range, the compression waves of the air will cause the ear drum to vibrate and that vibration will be heard by the owner of the ear as sound.

The speaker makes controlled changes to the pitch, loudness and harmonic content of the original vibrations which, if the hearer understands the language, will be interpreted as an intelligent message.

If a radio link is to be used to allow speaker and listener to be further apart, then the compression waves from the speaker's mouth vibrate a microphone diaphragm which produces an electrical waveform to be carried by the radio link. At the other end of the link, a similar electrical waveform vibrates the

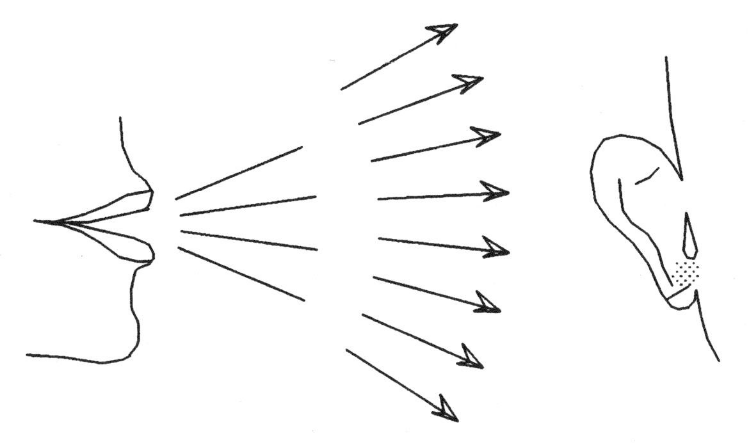

Fig. 4.1 *When a person speaks, waves of compression of the air radiate from the mouth*

diaphragm of a loud speaker or earphones to produce compression waves in the air close to the listener's ear (see Fig. 4.2).

This process will be successful if the electrical waveforms can be made a faithful copy of the original sound and the sound output of the loudspeaker a faithful copy of the electrical signals. The purpose of the radio link is to generate electrical signals at its output which are faithful reproductions of the electrical signals presented to it.

Note that the radio link makes no judgment about whether the electrical signals form an intelligent message or not. Its duty is served by reproducing the signal without significant distortion.

The electrical signal does not necessarily come from a microphone and go to a loudspeaker; other common sources of signal are computers or video cameras. In fact any signal that can be translated into electric form can be transmitted by radio.

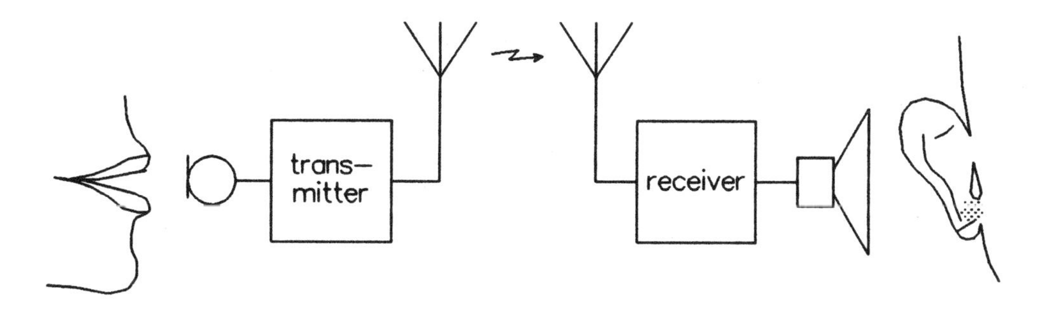

Fig. 4.2 *Use of a radio link to allow speaker and listener to be further apart*

4.2 KEEPING THE MESSAGE CLEAR

This section follows directly from Section 4.1.

The electrical output of a radio link will never be an exact copy of the input—there will always be some loss (parts of the original signal will be lost or moved in relation to other parts) and there will always be distortion (extra signals will be added that were not part of the original message).

The question is: How much of each form of loss or distortion can be tolerated before the intelligence of the message is affected? The answer to this question depends very much on the factors that are the keys to intelligence in each signal. If the signal is a voice, intelligence is transmitted by variations of frequency (heard as changes of pitch), loudness and harmonic content, so for a voice transmission, these are the factors that must be closely specified.

For communication of spoken messages, signals in the frequency range 300 Hz to 3000 Hz are required; the loud passages of the voice must be at least 10 times the power level of noise and interference (which may be written as '10 decibel signal-to-noise ratio'); and distortion which produces extra harmonic content will affect intelligibility if it is more than about 5%.

For high-fidelity transmission for entertainment, the signal must be kept much clearer. The frequency range must be from 50 Hz to at least 10 kHz; signal to noise ratio must be at least 40 dB (10 000 times the power) and preferably 50 to 60 dB; and harmonic distortion must be no more than a fraction of 1%.

For more information on decibels and the general subject of power measurement, see Section 10.10, 'The decibel scale'.

If the signal is a computer program, the logic states are simply switched on or off and intelligence is transmitted by the timing of the switching. Harmonic distortion is almost irrelevant, and frequency response can be accommodated within wide limits. The factors that are important are minute variations of the total time delay in the system (which are described as *jitter*); signal-to-noise ratio; and the effect of occasional deep fades which may cause loss of a few bits of the program information.

A television program has its own set of important considerations. One, for instance, is a factor called *group delay* which measures the total time required for transmission of the high-frequency components compared with the time for transmission of lower frequencies. Errors of only a few nanoseconds will show on the received picture as a smearing of vertical lines.

For a voice transmission, group delay errors several thousand times as big would not even be noticed. On a computer program, group delay may be a consideration but not quite to the critical extent that it controls the quality of the picture in a television system.

There are many forms of degradation of the signal that can affect television, and different methods of transmission of the signal can be more sensitive to some of these considerations than to others.

When an intelligent signal is modulated onto a radio carrier wave, the process is often very similar to the processes which cause distortion. Care is needed in design and operation to ensure that the signal transmitted is actually intelligent.

4.3 METHODS OF MODULATION

There are no prerequisites for this section.

On the surface there appears to be a whole range of ways of modulating a signal onto a radio carrier—the following list gives a selection of examples of names and acronyms:

AM amplitude modulation
FM frequency modulation
SSB single side band
PM phase modulation
CW continuous wave (telegraphy)
PCM pulse code modulation
vestigial sideband
BMAC type 'B' multiplexed analogue components

and a host of others for specialised purposes.

Despite all the apparent differences and different names, all forms of modulation can be revealed as one or other of two types:

1. *Amplitude modulation* covers all processes where the short-term output power of the transmitter is controlled by the signal.
2. *Frequency modulation* includes all cases where the output power is kept constant and the frequency is varied over a small range under control of the signal.

The next few pages contain some explanation of how each of these works and what some of the various names mean.

4.4 PRACTICAL AMPLITUDE MODULATION

Section 4.3 is a prerequisite for this section.

One method that illustrates the principle is to arrange for the output stage of the radiofrequency generation section of the transmitter to be running at constant efficiency and then to vary the power input to it under control of the program (Fig. 4.3).

There are difficulties associated with this simple arrangement. One difficulty is that for an audio program (speech and music), the controlling device must be a high-fidelity audio amplifier with power output of half the total power being controlled. Thus for a 100 kilowatt broadcasting transmitter in which the final stage is running at 70% efficiency, the total power to be controlled is 143 kilowatts and the audio signal required is 71.5 kilowatts of high-fidelity audio power.

The audio amplifier must run in a linear class of operation which indicates lower efficiency of power conversion so it may draw more power overall than the radio frequency section.

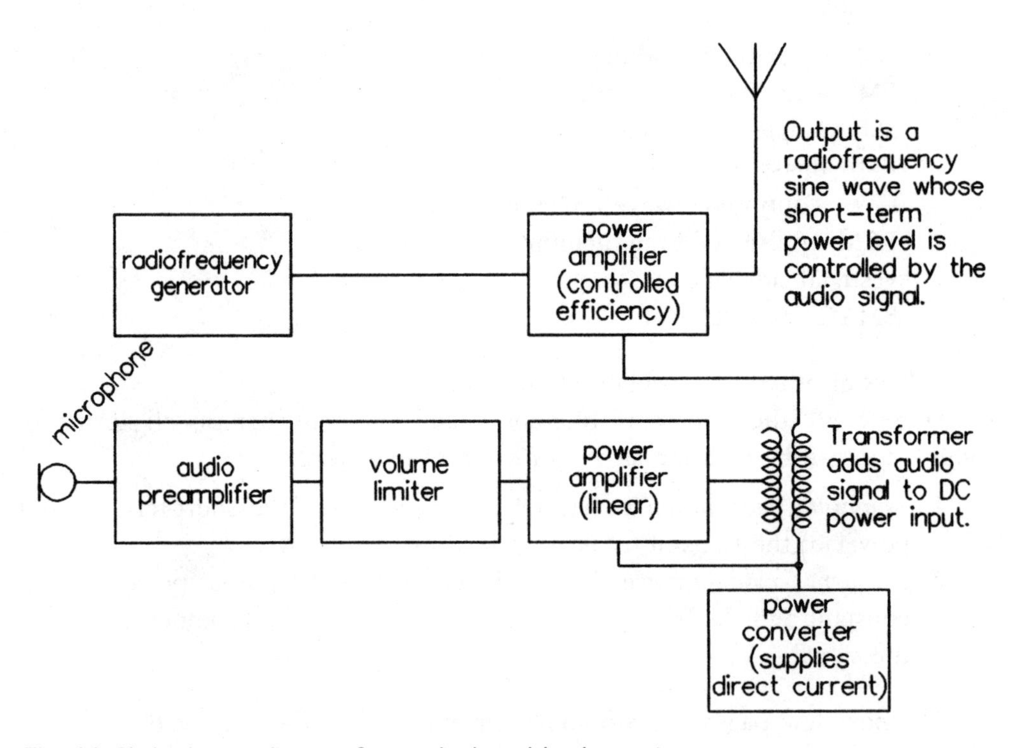

Fig. 4.3 *Block schematic diagram of an amplitude-modulated transmitter*

There are also some points in the circuit where the peak voltage is very high, in theory up to four times the voltage delivered by the power supply. For high-power valve amplifiers the DC supply may be in the range of 10 000 to 15 000 volts, so peak voltages up to 60 kv may need to be designed for.

Over the years there have been many attempts to sidestep the problems associated with this form of modulation, but in each case the different system has been found to have other defects which are more serious than those mentioned above. The result is that, at the present time, most manufacturers of high-power transmitters have accepted the defects and used circuits based on the simple system.

The big advantage of amplitude modulation is that the receiver is simple, the detector can be a diode as used in a crystal set (the simplest of all receivers), and no fussy adjustments are needed to get it working.

In the case where one transmitter broadcasts to many thousands of receivers, the total use of resources is minimised by simple receivers so amplitude modulation is used for broadcasting on the medium-frequency and high-frequency (short-wave) bands.

4.5 PRACTICAL FREQUENCY MODULATION

Section 4.3 is a prerequisite for this section.

The aim of a frequency-modulation (FM) system is to keep the output power of the transmitter constant and vary the frequency of the carrier by a small amount in step with the intelligent program. Figure 4.4 shows the waveform of a frequency-modulated signal.

Variation of the frequency must be done in the radiofrequency oscillator which is the first stage of the radiofrequency section. The circuit used acts on one of the tuning components in the oscillator to vary that component's effect under control of a voltage which includes the program to be transmitted. Circuits of this type are called *reactance modulators*.

For transmitters, FM has several technical advantages over amplitude modulation. There is no need for a high-power high-fidelity audio amplifier and

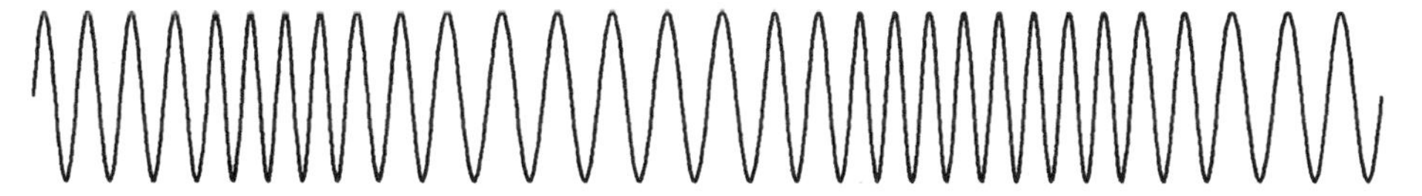

Fig. 4.4 *Frequency-modulated waveform*

no need to cater for very high voltages. The efficiency of the power output stage does not need to be kept constant over a range of supply voltages, so it can be driven to maximum efficiency.

For receivers, a well-designed FM system has a capture effect. Provided that a signal can be found at all, a given level of transmitter power over a given path loss will give better signal-to-noise ratio and be less subject to interference if the link uses frequency rather than amplitude modulation.

However, when signal strength fades below a threshold level, the capture effect ensures that the signal is completely lost, and if the interference ever becomes stronger than the wanted signal, then the interference takes over and the wanted signal is lost.

Frequency modulation is ideally suited for transmission of high-fidelity sound in conditions where a strong, clear signal can be provided, for example broadcasting over an area the size of a city and its immediate surroundings.

Frequency modulation requires wider bandwidth than amplitude modulation for similar performance specifications. FM generates sidebands which are not so easy to analyse as for AM, but as a rough idea imagine two AM signals side by side in the spectrum working so that the timing of minimum power of one corresponds with the maximum power of the other. The total power of the two combined stays constant but the observed carrier frequency shifts back and forth between the two signals as is required for FM.

A full analysis of the sidebands of an FM system requires use of a mathematical tool called *Bessel functions*; however, a full analysis is not often needed.

The name *deviation* is given to the cyclic variation of carrier frequency in response to the modulating signal. Deviation is small for low levels of modulation such as soft passages of music, and is greater for high levels such as shouted messages.

Modulation performance of an FM system is checked by measuring deviation and checking it against the input signal.

The bandwidth required can be closely estimated by taking the total excursion of frequency (twice the nominal figure for peak deviation) and adding to it the figure for highest modulating frequency to be transmitted. For example, the VHF broadcasting standard calls for peak deviation of 75 kHz and maximum modulating (audio) frequency of 15 kHz:

$$2 \times 75 = 150$$
$$150 + 15 = 165$$

The 165 kHz bandwidth required fits neatly in the 200 kHz channel spacing of the band.

The wide bandwidth requirement means that only the VHF, UHF and higher frequency bands have room in their spectrum for FM signals. Present administration does not accommodate FM on carrier frequencies below 30 megahertz.

4.6 SINGLE SIDEBAND (SSB)

To follow the explanation in this section you will need to understand Section 2.7, 'Sidebands', and Sections 4.3 and 4.4 in this chapter.

The principle of SSB is a variation of amplitude modulation. The AM signal consists of a carrier wave and two sidebands which carry the intelligent message. From the point of view of getting the message from place to place, the carrier does nothing and the full message is contained in either one of the sidebands. Therefore the full message requires only one sideband on its own. Figure 4.5 shows the spectrum of an SSB signal.

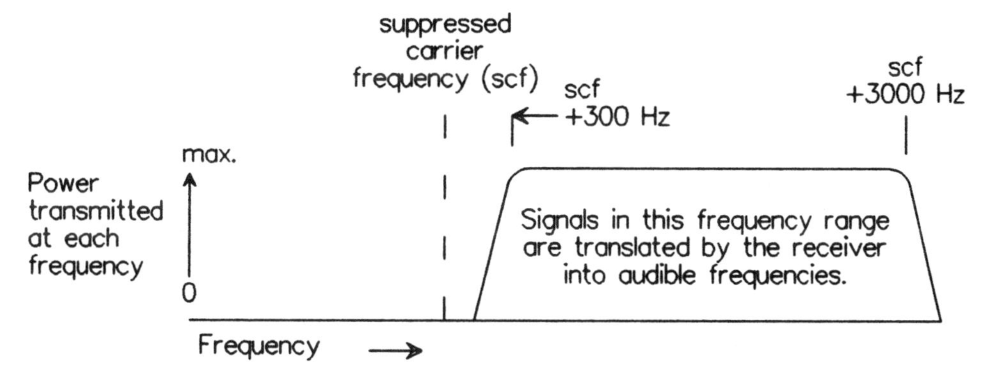

Fig. 4.5(a) *Spectrum of an SSB signal*

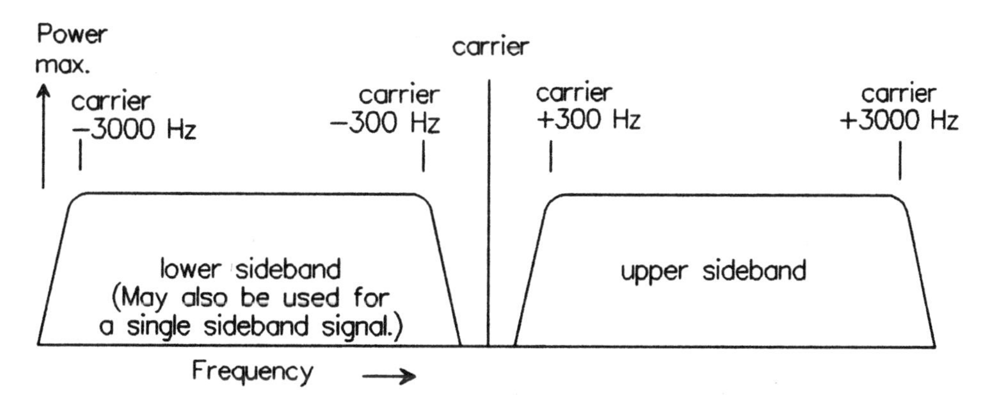

Fig. 4.5(b) *Spectrum of an AM signal shown for comparison*

The significance of a spectrum diagram such as Figure 4.5 may not be immediately obvious. If you have difficulty understanding what the diagram means think of a very narrow band receiver being slowly tuned over the signal and making a measurement of average power at each step of the tuning. The *x* axis shows the position of the tuning dial of the receiver and the *y* axis shows the average power measured at each step of the tuning. The primary aim of Figure 4.5 is to show why it is that when compared with an SSB signal the amplitude-modulated signal takes up over twice as much spectrum space (you would have to turn the receiver dial twice as much) and is a less efficient user of transmitter power (it has most of the power concentrated in the carrier which contains no intelligence).

This has advantages for the transmitter in that there is a saving of power, and for the receiver in that all the power used is concentrated in the message; the 'talk power' of the signal is raised. Also, because noise tends to be fairly evenly spread across the spectrum, the signal-to-noise ratio can be improved simply by making the bandwidth of the receiver less.

From the spectrum administration point of view, less bandwidth allows twice as many users in the same space, and because no carriers are radiated, there are no steady whistles of heterodynes in the background noise.

Although the principle of SSB is simple and the advantages obvious, the realisation of a workable system has taken many years of technical refinement with many small improvements by many different people.

The critical factors are: a filter which has a passband flat over the range of the sidebands generated by voice frequencies and with very sharply tuned sides; and in the receiver the carrier must be replaced by a locally generated signal (carrier insertion oscillator) of exactly the same frequency to within a few hertz.

In modern SSB transceivers, the signal is usually generated at a common low frequency by the combined action of a balanced modulator and a filter circuit and then translated to the working frequency by heterodyning in a second balanced modulator. The circuitry of the receiver is very similar with functions performed in the opposite order. Because of that, the SSB principle has found most use in two-way transceivers where the same components can be used for both transmit and receive, with configuration changed with switching.

SSB is the method of modulation which can transmit a voice with least use of spectrum space so it has become the standard for mobile communication systems in the HF band.

There have been some experiments with single sideband use for broadcasting in the international short-wave bands which have used a reduced, rather than a completely suppressed, carrier to avoid the need for a highly accurate source of carrier reinsertion signal. Results have not so far been encouraging because it is difficult to make fidelity of the received audio output up to accepted broadcasting standards.

ISB (independent sideband) transmission is an advancement of the SSB principle where both sidebands are generated and received as two separate communication channels. Systems of this type are used when two channels of, for instance, telephone communication are needed to a temporary camp in a remote area; they may also be used by the defence forces who may need to rapidly establish communication links independent of wired systems.

Television broadcasting uses an adaptation of the SSB principle called *vestigial sideband*. The purpose is to minimise the usage of spectrum by the very broadband signal (components up to nearly 4 MHz) but to do it in a way that does not require the fussy adjustment of carrier reinsertion.

4.7 PHASE MODULATION

Sections 4.3 and 4.5 are prerequisites for this section.

Phase modulation (PM) is a variation of the principle of frequency modulation. The same reactance modulator circuit is used on one of the tuning components, but in this case it is in a later stage in the radio frequency circuit.

The essential difference between frequency modulation and phase modulation is that phase modulation is not able to transmit a steady shift of frequency to indicate a DC condition.

For voice transmissions heard by ear, phase modulation is almost indistinguishable from frequency modulation with a pre-emphasis frequency response characteristic.

In mobile communications transceivers controlled by crystal oscillators operating in the relevant bands between 35 and 512 MHz, the transmit section is often a phase modulator and the receiver a frequency detector with an appropriate de-emphasis to correct the audio response.

The same measurement of modulation performance using a deviation meter as is done with frequency modulation can be used with PM provided that the difference in frequency response is allowed for.

The term *angle modulation* is sometimes used by regulating authorities to include both frequency modulation and phase modulation in those documents intended to apply equally to both.

4.8 PULSE CODE MODULATION (PCM) AND CONTINUOUS WAVE (CW)

Section 4.3 is a prerequisite for this section.

Pulse transmission is a general term to describe all conditions where the transmitter operates in only one of two states—either steady transmission of maximum power, or off.

A ship's radar unit is an example of simple pulse transmission. The transmitter makes an extremely short burst of signal at a multiple kilowatt output level and then switches to receive to wait for an echo. The waiting time is thousands of times longer than the transmit time, so average power level is only a very few watts. Figure 4.6 shows the signals in a ship's radar unit.

In a communication system, the pulses may be modulated to carry an intelligent signal. It is possible to vary any of the time-related factors, that is, the repetition rate, the width or the duty cycle to carry a message. Note once again that the radio system itself makes no judgment about what constitutes an intelligent message; it is up to the system designer to decide on protocols to give meaning to the modulation.

CW telegraphy (continuous wave) is a particular form of *pulse code modulation* (PCM). The name is a historical relic and these days may be a bit misleading. The first radio transmitters were sparks like tame lightning bolts which intrinsically give a signal over a wide range of frequencies. This rough signal was fed through a tuned circuit straight to the radiating system. In the earphones of a receiver, it gave a buzzing sound.

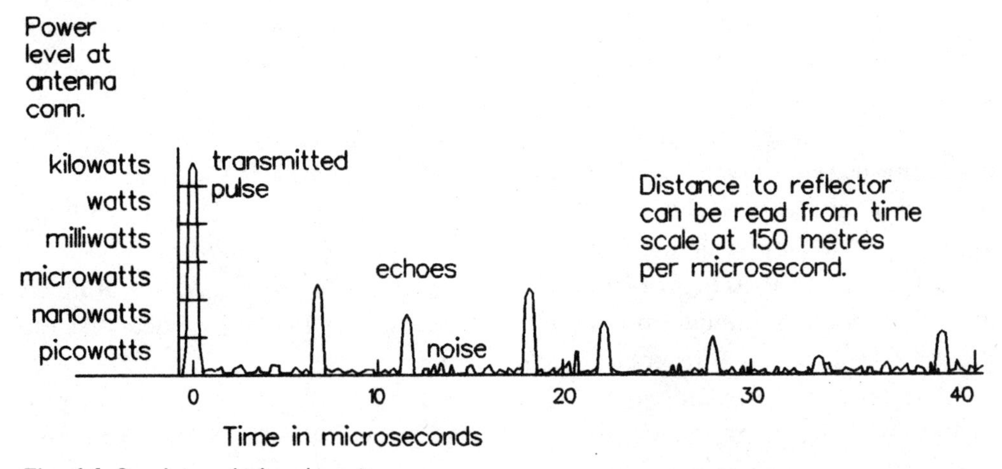

Fig. 4.6 *Signals in a ship's radar unit*

Later transmitters used a signal generated on a single frequency by a valve oscillator and then amplified to the required output level by other stages of the radio frequency circuit. In the earphones of, for instance, a crystal set, this signal gave no steady tone; there would be a click at switch on and another click at switch off. These transmissions could be made audible by arranging for some signal from a local oscillator to be added. The local oscillator, tuned to anywhere between 500 Hz to about 2 kHz different from the wanted signal, would reveal it as a steady musical note.

The name *continuous wave* was given to these single-frequency transmissions originally to distinguish them from spark transmissions. When the sparks 'died out', the name stuck and now 'CW' means any transmission where the carrier wave is simply switched on or off under the control of a hand key.

Originally, both spark and CW transmitters used the Morse code or one of the related telegraphic codes derived from Morse. These telegraphic codes are a particular form of pulse code modulation (see Fig. 4.7) which is designed for generation by a human hand on a key and for decoding by a human ear.

Machines (such as computers) and telegraphic codes do not mix very well. Computer generation of Morse code is no great problem but decoding is. The fact that characters do not all have the same length presents a difficulty for the machine to decide when the present character really has finished.

For everyday communication between computers (and related machinery), codes where the end of a character is easily defined are much more reliable. One of the more common is ASCII (American Standard Code for Information Interchange). In this code there is a defined starting signal for each character and a defined stop signal. There are 7 bits carrying information which can signal

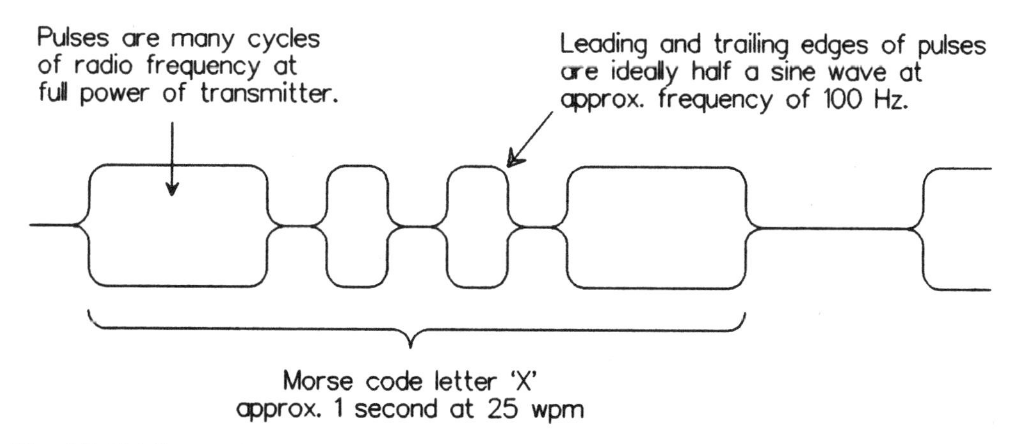

Fig. 4.7 *CW Morse code transmission*

128 separate characters, and a parity bit for checking true transmission of the information. A machine receiving this signal can do so with much lower error rate than for a telegraphic code, and can signal when an error is detected, which may be more important in some cases.

CW transmitters hand-keyed using a code related to Morse and detected by human ears have a place and always will in the case of emergency situations. In adverse conditions, such as standby transmitters operating on emergency power with a jury-rigged aerial and not able to wait for the best time of day, a human operator can detect and follow the message of a faint heterodyne whistle in conditions where a machine could not even be made aware that a signal exists. Compared with voice transmission, the code signal can be easily fitted within 100 Hz bandwidth which immediately gives it a significant advantage in noisy conditions, and the need only to recognise a tone rather than understand a voice makes another step of advantage for the code signal.

Thus a CW transmission to a trained human listener will, for the foreseeable future, be the means of transfer of information with best possible intelligibility under the worst of adverse conditions.

4.9 MODULATION FOR COMPUTER DATA

This section follows directly from Section 4.8.

Digital data cannot be fed straight from a computer to a radio communication transceiver designed for transmission of voice signals. The data may be a continuous series of logic '1s' which would appear to the radio system as a DC input at the logic 1 level; or it could be a stream of logic '0s' which also appears as a DC level; or it could be alternate '1s' and '0s' which would be presented to the radio as a square wave.

Voice communication systems cannot transmit a DC level at all and are not particularly good at transmitting square waves cleanly. Wired telephone systems have exactly the same incompatibility and the problem is solved with a modem—the data is frequency modulated onto a tone in such a way that DC levels of the data signal can be assigned to particular tone frequencies. The same principle can be used for radio transmission of data—the modulated tone from the modem is accepted by the radio in place of a voice signal.

Radio modems are used in all places where the communication channel must carry both speech and data. Where a radio link is established specifically to carry data and it is known in advance that no voice communication will ever be required, a transceiver of a different type can be used. This transceiver uses the same principle of modulation as the modem but works directly on the radio frequency carrier—DC levels can be directly assigned a corresponding frequency in the output passband of the radio link.

These directly modulated systems offer an advantage in information density. The radio modem system generates sidebands for the tone and also sidebands to signal changes to the tones. In directly modulated systems, the carrier takes the place of the tone and only the sidebands for changes are needed. The practical limitation on speed for a modem system using one pair of logic states (for 1 VF channel) is about 2400 bits per second (b/s); for the same use of spectrum space, the directly modulated system can work at 4800 b/s. If multiple logic states can be used, that speed can be doubled or sometimes quadrupled.

4.10 THE 1 VF CHANNEL

Section 1.6, 'Frequency', is a prerequisite for this section.

Our ears actually hear vibrations in the air in all the frequency range between about 16 hertz and 16 000 hertz (that is, for those of us whose ears are still young enough to work with full efficiency).

One hertz is a frequency of one cycle of vibration per second; for instance, the phrase 'a continuous tone of 1000 hertz' is actually describing a series of vibrations, each of which takes one-thousandth of a second to pass.

In many experiments conducted over many years, it has been found that different parts of the spectrum of sounds we hear are used by our ears for different functions. A quite large range of the highest frequencies is used only to add timbre to voices and to identify the direction of the source of a sound. The intelligence part of a message is concentrated in the mid-range with frequencies between about 300 and 3000 Hz (often written as 3 kHz).

Many noise sources spread their energy fairly evenly over the frequency spectrum which often means that most of what is heard as noise and interference appears to be concentrated in the higher pitched range of sounds. For this and other reasons, it has become normal to transmit for communications purposes sounds only in the mid-frequency range (approx. 300 Hz to 3 kHz) and quite severely filter other frequencies.

There is minor variation in the actual limit frequencies for different classes of service. Where background noise is most severe, as for instance when conditions are bad on an HF radio service, the bandwidth may be restricted to as little as 2.1 kilohertz for maximum signal-to-noise ratio. In the public telephone service, widening the bandwidth to 3.4 kHz has been found to give a useful increase to a person's ability to recognise individual voices.

This range of frequencies between about 300 Hz and 3 kHz, which is most useful for communication purposes, has been named 'the voice frequency channel'.

4.11 MODULATION PERCENTAGE

This section is an extension of the concepts of 'Sidebands' (Section 2.7) and 'Occupied Bandwidth' (Section 2.8).

The three-tone pattern shown in Figure 2.6 is a graph of the waveform of an amplitude-modulated radio signal. The two side frequencies must be exactly half of the amplitude (in terms of voltage or current) of the centre frequency and the amplitude of the resulting pattern is just reduced to zero at the bottom of the trough. This special condition is called *100% modulation*.

If the amplitude of the sidebands is less than what is required for that special condition, and phase and frequency relationships are still correct, then a similar pattern is produced but the peak does not go as high and the trough does not go to zero. Numbers between 0 and 100 are used in a percentage scale to express any desired depth of modulation between unmodulated carrier and just complete cancellation at the bottom of the trough (see Fig. 4.8).

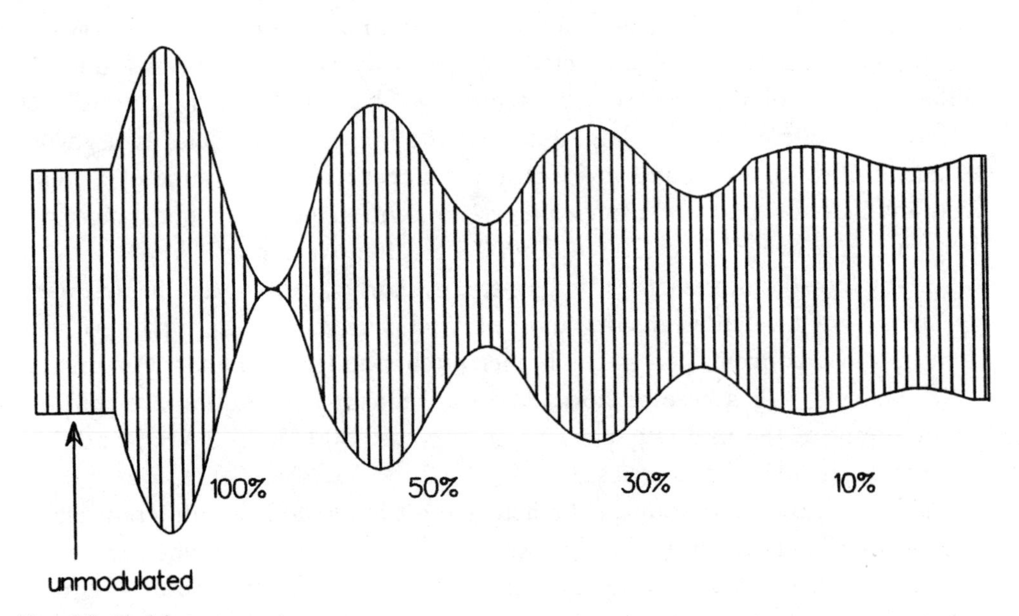

Fig. 4.8 *Modulation percentages*

Percentage scales for modulation depth are also used for other forms of modulation, but in some cases the state of 100% modulation is more arbitrary. For frequency modulation, for instance, 100% corresponds to the maximum deviation that can be used without causing adjacent channel interference.

4.12 OVERMODULATION

This section follows on directly from Section 4.11.

If the amplitude of the sidebands is greater than half the amplitude of the carrier, it should in theory be possible to generate a sensible modulation envelope and transmit it. In the practical case, however, there is a limitation of the electronics of the transmitter—it is not possible by the 'power control' method of amplitude modulation to reduce the output to less than zero.

The result of attempting to modulate over 100% is distortion of the modulation envelope as shown in Figure 4.9.

The effect of this distortion is exactly the same as if the carrier were modulated with a distorted waveform. Distortion introduces extra components of the modulating signal, and those components appear as extra sidebands at the harmonic frequencies of the original. These harmonic frequencies raise the bandwidth of the original message, and in the same way, distortion of the modulation envelope causes extra frequencies in the sidebands which widen the occupied bandwidth of the signal. Extra modulation components interfere with adjacent channels causing clicking, popping and hissing noises. This form of interference is called *splatter*.

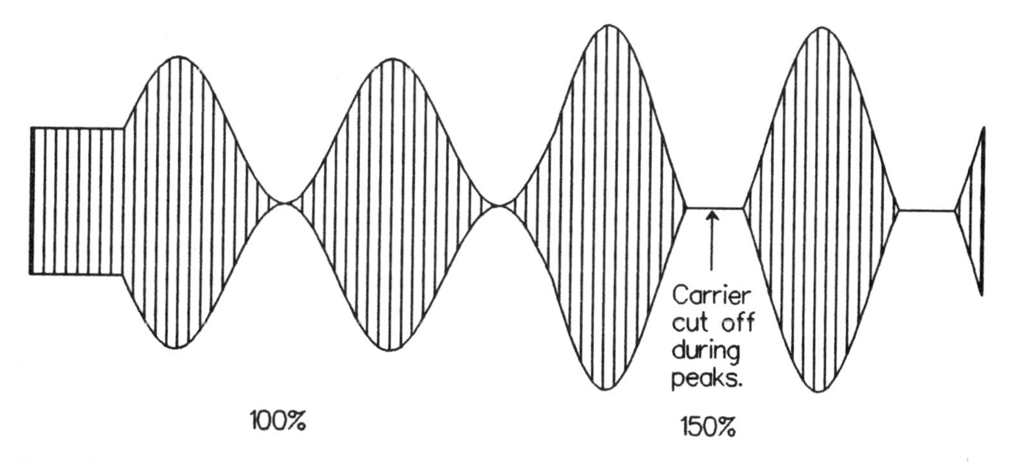

Fig. 4.9 *Overmodulated AM signal*

With the angle methods of modulation, the observed effect of overmodulation is very similar, but for a different reason. Greater modulation causes greater swing of frequency and higher modulating frequency (that is high-pitched sounds in the program) causes the sidebands to be spread wider. Programs most likely to cause adjacent channel interference of an FM signal are sibilants and high-volume passages in the treble register.

4.13 CLIPPING AND LIMITING

Voices and sound programs are very variable in the volume of sound they present to a microphone. Even monotonous-sounding voices have peaks of volume that are 6 to 10 dB (decibels) above the average level of sound (4 to 10 times the power). For voices that are well articulated in the theatrical sense, this ratio between peak and average may be as high as 20 dB (100 times the power). The peaks of volume last for less than a second each and come at the rate of about 1 to 3 per minute. The electronics of the radio system must not allow the peak level to cause overmodulation, so the result is that average modulation level for a signal of that type is very low.

Circuits are available which reduce the level of these rare peaks relative to the average program level. If used in moderation, they are able to make the average about 6 to 10 dB closer to the peak level with very little change in the overall quality of the transmitted voice. The modulator gain control is set in relation to the peak level so this can have the same effect as a 4 to 10 times increase in transmitter output power.

One type of circuit called a *clipper* simply chops the peak off, refusing to respond over a certain level no matter how big the peak is. This function by itself is the same as overmodulation but the clipper is followed by a passband filter to remove the resulting harmonic frequencies which cause adjacent channel interference. (See Fig. 4.10.)

The *limiter* circuit senses the peak arriving and rapidly adjusts the system gain for a short time (¼ to ½ a second) while the peak passes. This preserves the waveshape of the peak. (See Fig. 4.11.)

The two diagrams in this section show the effect of each piece of equipment. If each cycle shown were taken to represent about 50 to 100 actual cycles of an audio program, they would sound fairly close to a syllable made of the letters 'PT' run together. At this level of restriction, the clipper will noticeably soften the explosive parts of the sound; the limiter will preserve the shape of the syllable but reduce it as a whole in comparison to other passages of the program.

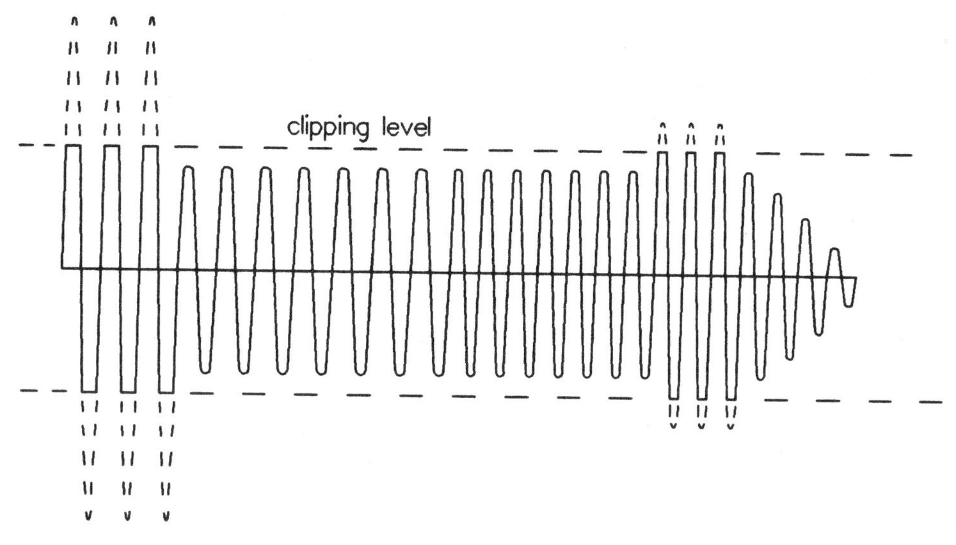

Fig. 4.10 *A clipped syllable*

Clippers are capable of raising the 'talk power' of a voice more than are limiters but they do cause some distortion of the audio signal. In systems where the intelligibility of the message is paramount, a clipper is able to raise the average level to within a couple of decibels of the peak with only moderate cost in distortion. Clippers are most appropriate in communication systems that

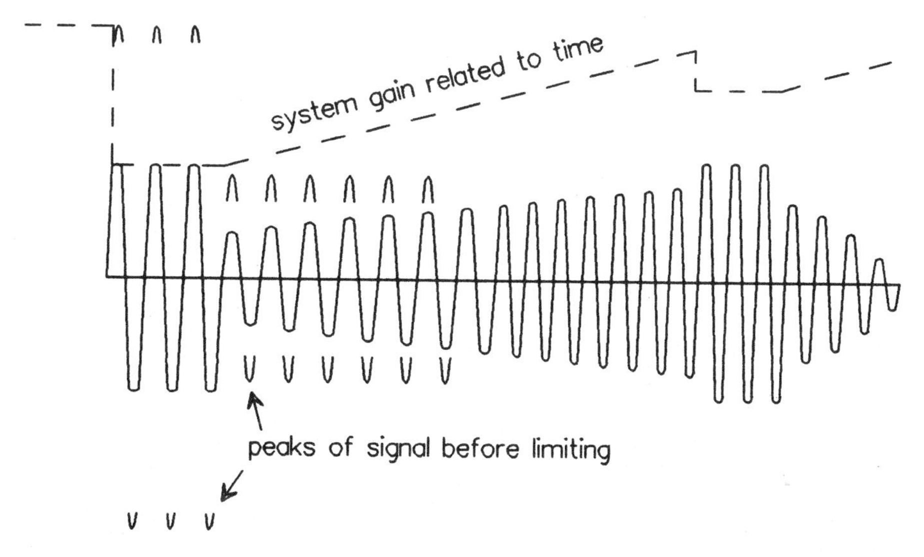

Fig. 4.11 *A limited syllable*

transmit spoken messages. Limiters are most appropriate for entertainment systems which include such things as orchestral music.

Broadcasting transmitter stations must include an effective means of preventing overmodulation somewhere in the chain of program input equipment. Either a clipper or limiter will do that job.

CHAPTER 5

HOW SIGNALS GET THERE

This chapter builds on the principles outlined in Chapter 3, 'How energy is coupled'. In general, an understanding of Chapter 3 is a prerequisite for all of this chapter. It also presumes you have some basic understanding of the subjects covered by daily published weather maps.

5.1 PROPAGATION IN FREE SPACE

If there were absolutely nothing in the way from transmitter to receiver, you could make a simple calculation of signal strength by using the 'inside of a hollow sphere' model mentioned in Chapter 3, Section 3.4. The laws of optics, and in particular the 'inverse square law', would apply to all radio frequencies in this simple case. To make a close estimate of received signal strength you would need to take into account only the path distance, frequency of the signal, and directional properties of transmit and receive aerials. This simple situation actually applies to most uses of radio in space research and communications. (For a diagram of the principle of the inverse square law, refer to Fig. 3.8.)

For most terrestrial communications, however, there is something in the way of the direct path; in terms of optics the light must shine around a corner.

There are a few places where the 'free space' conditions apply on the Earth's surface. If the two ends of the link are close enough to each other and high enough to be able to see each other, and the combination of wavelength and tower height places the radiating elements several wavelengths above the nearest obstructions, then the 'free space path loss' sums will give a close estimate of received signal strength. That part of the propagation mechanism in which signals follow a direct path and free space path loss conditions apply may sometimes be known as the *space wave.*

For radio paths over an obstruction, the calculation of a 'free space' signal strength represents a theoretical maximum; the obstruction always weakens the signal and all calculations of signal strength are to show the degree of that extra loss.

5.2 THE MECHANISMS OF PROPAGATION OVER OBSTACLES

Radio signals obey the basic laws of optics. There are some differences, however, compared with the application of those laws to light beams. (Calculations of the degree of each effect must allow for the vastly different wavelengths. A half-wave dipole aerial is the finest possible point source; generally the image-forming properties of a lens do not apply.) With allowance for those differences, the effects of refraction, reflection and diffraction can be clearly identified and calculated according to the laws of optics. Figure 5.1 shows the mechanisms of the propagation of radio signals over an obstacle.

Libraries carry books on optics with reference numbers in the range 535 to 535.4. This section of references also includes some on application of optical principles to other parts of the electromagnetic spectrum.

There are also some mechanisms which are peculiar to particular ranges of frequencies within the radio bands and some that are specific to particular weather conditions or particular types of location.

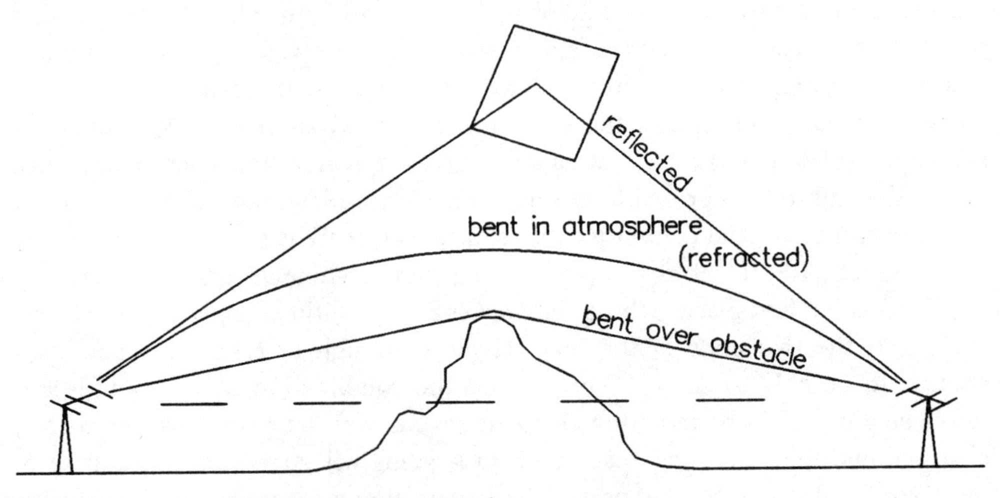

Fig. 5.1 *Propagation of radio signals over an obstacle*

One branch of optics that has much more effect on radio signals than on visible light is *Huygens's principle*. The part of his work that is relevant is the theorem that the direction of propagation of the wave is always at right angles to the wavefront. Any factor which slows the wavefront in a local area will cause the direction of propagation to bend.

5.3 WHY SIGNALS FADE

The practical application of signal strength calculations must always include a margin to allow for fading. Fading usually results from having two or more signals arrive at the receiver input by different paths. Relatively minute differences in path length can cause the two signals to cancel each other. The mechanism that causes this effect is essentially the same as that which causes the two-tone beat pattern displayed in Figure 2.5.

In cases where the path difference depends on weather conditions or where one of the stations is mobile, the signals as you hear them in a receiver can go through a fairly regular cycle of cancellation and reinforcement over a range of signal strengths amounting to about 40 decibels. Figure 5.2 shows a graph of signal strength resulting from two equal signals steadily drifting in phase. The reference line shows the signal strength of either signal on its own. When the two signals are equal in strength and exactly in phase (see Section 7.8 for explanation of 'phase'), the received signal is 6 decibels greater than either one.

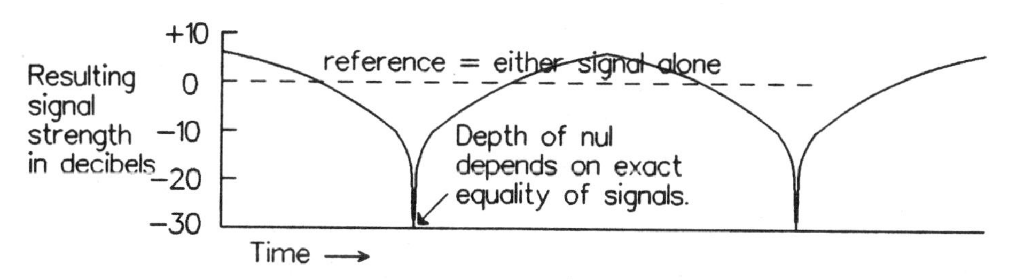

Fig. 5.2 *Graph of signal strength resulting from two equal signals steadily drifting in phase*

If the same two signals are exactly in opposite phase, they cancel and signal strength is theoretically zero.

The next section will give a short description of each of the major mechanisms of propagation. In the next chapter there is a list of frequency bands in the radio spectrum with information on which mechanisms have most effect on them.

Note that the whole of this chapter is an outline only. Propagation of radio signals is a subject on which quite thick books have been written, and those thick books still do not give exhaustive detail of all possible situations. Books on the subject are listed in libraries under the classification number 621.384.

5.4 GROUND WAVE AND DIFFRACTION

Ground wave and diffraction are two factors that are quite different in their mechanism, but in many cases both have effects on the signal which display the same characteristics.

Diffraction works on all frequencies (radio, light, X-rays etc.) and is the mechanism that causes slightly fuzzy edges on shadows. It is an effect that has been well researched in the optical field so if you are particularly interested in it, books on the subject of optics are a good source of information.

When a radio wavefront travels over the Earth's surface (including the watery parts), the wave induces electric currents in the surface layers of soil or water which tend to travel with the wavefront but slightly more slowly. The earth currents have the effect of slowing those parts of the wavefront which are near the surface, and that effect in turn causes the lower parts of the wavefront to be curved backwards.

A demonstration of the principle can be shown by holding a sheet of material such as a curtain or bed sheet so that it hangs with one edge in very light contact with the floor and then slowly moving it sideways. The sheet demonstrates what happens to a plane wavefront as it moves over a partly conductive surface. The top part of the sheet will hang vertically but at the bottom, the sheet will curve back due to the drag of the floor (see Fig. 5.3).

Due to Huygens's principle, the part of the wavefront near the surface propagates in a direction tending towards the surface. In the case of a radio wave near the surface of the curved Earth, the practical effect is that the energy of that part of the signal tends to follow the curve of the Earth.

When a mobile station is moving over the Earth's surface in a direct line away from a fixed transmitter (ships at sea for instance), a distance is reached where the mobile aerial crosses the geometric shadow line. The signal does not suddenly fail

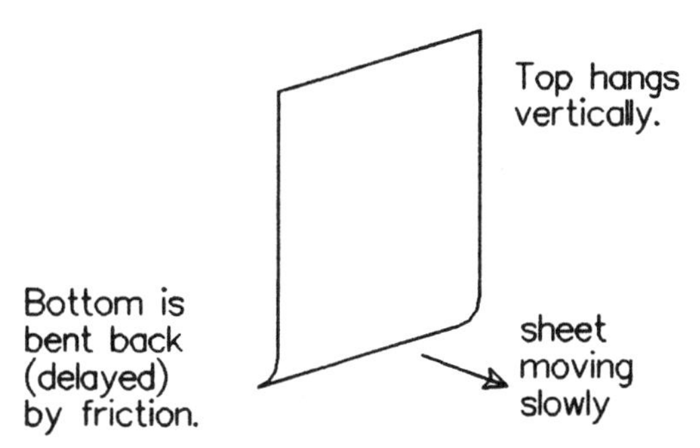

Fig. 5.3 *Demonstration using a sheet to show a radio wavefront travelling over the Earth's surface*

at that point but from then on gradually becomes weaker in a regular fashion over a distance which may be a few kilometres for a VHF signal ranging up to thousands of kilometres for a signal in the VLF band. (See Fig. 5.4.)

For ground-wave/diffraction signals travelling over the Earth's surface, it is possible to specify a particular rate of loss in terms of decibels per kilometre of travel. That figure depends on the frequency being tested and on the conductivity of the ground. The lowest frequencies give the lowest figures of loss. Ground-wave effects can be seen for all radiofrequencies but are relevant only for practical

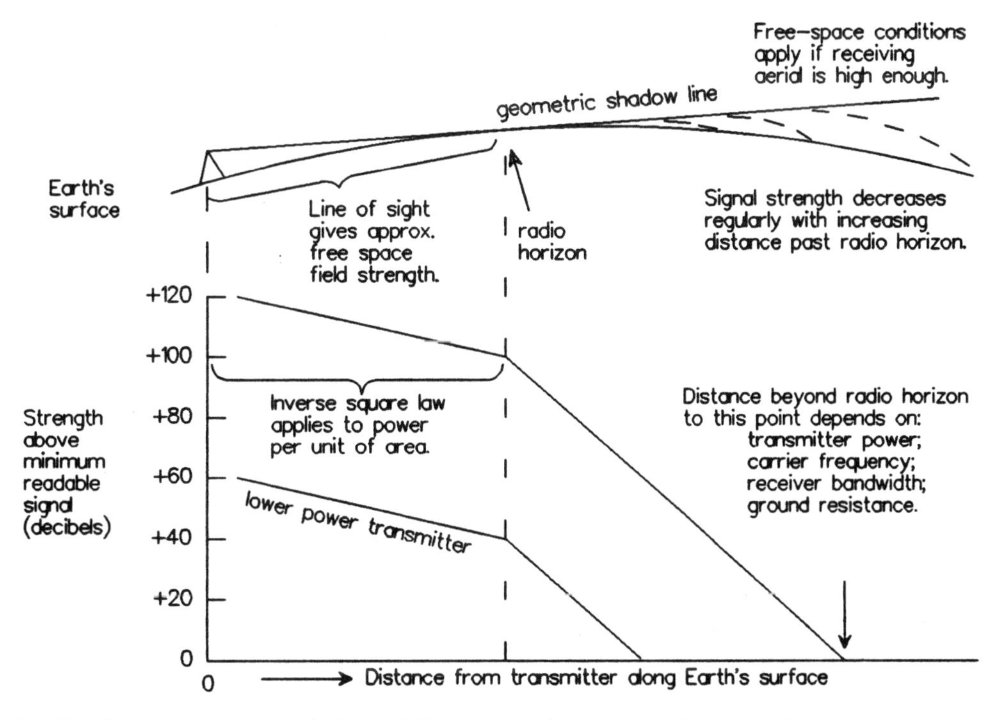

Fig. 5.4 *Expected signal strength due to diffraction/ground wave (general shape only)*

calculations for frequencies below about 50 MHz. In the UHF band and higher, the ground-wave signal would be completely lost within a kilometre or so.

For good transmission of ground waves, the surface should either be a perfect conductor or a perfect insulator of electricity. Sea water is noted as the best and some of the soils where clay is mixed with haematite are among those with the worst reputation. Pure dry sand is an insulator of electricity and therefore would be transparent to radio waves; however, a slight trace of dampness mixed with some salt would render a sandhill lossy to radio signals.

For field strength measurements at ground level, it is impossible to distinguish between the effects of diffraction and ground-wave propagation. Careful measurements related to altitude at ranges well beyond the geometric line of sight may show a distinction between the two effects.

Pilots of light aircraft using non-directional beacons (low-frequency end of the MF band) for direction-finding occasionally observe that they can get a workable signal while parked on the runway of a bush airstrip but lose the signal for a short time just after take-off. The aircraft on the runway hears the signal by ground wave which rapidly becomes weaker as the craft takes off and climbs; after more climbing the effect of diffraction is heard and signal strength rises towards the free-space figure. (See Fig. 5.5.)

5.5 REFRACTION

For radio waves, the mechanism of refraction is the same as that investigated by high school students using blocks of glass and rows of pins. The practical effect, however, may show up in ways that would not easily be recognised by comparison with the experiment. At the scale of measurements for radio wavelengths, the

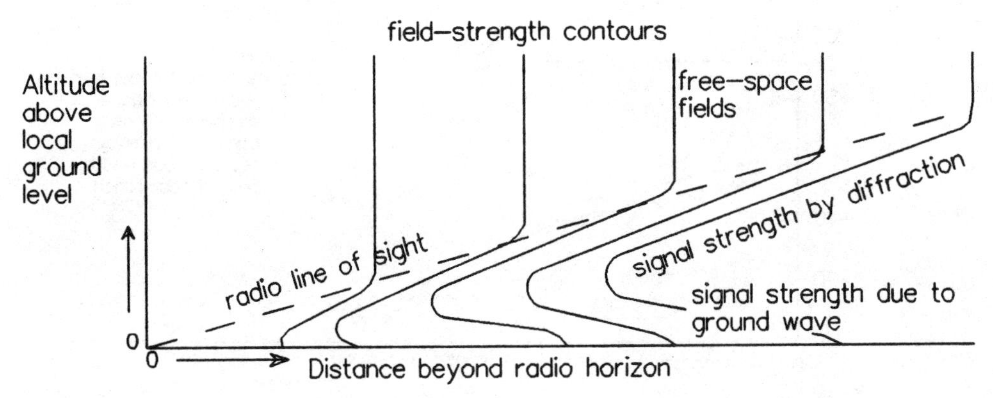

Fig. 5.5 *Graph showing separate effects of diffraction and ground waves (general shape)*

principle is more easily understood as follows: 'any object or medium which changes the speed of propagation of a radio wave will cause a deflection in the wavefront'. This, by Huygens's principle, will alter the direction of travel.

Layers of air in the lower atmosphere commonly have a refractive effect on some frequency bands. The layer of air at sea level is usually denser and more humid than the layers above it, and these factors slow radio waves. There is thus a normal tendency for radio waves to bend towards the Earth as they travel. The effect observed by a user is that the horizon for radio waves is usually further away than the geometric horizon. In average weather conditions, the ratio of radio horizon to geometric horizon is about 4:3. On a diagram, the ray path of a radio signal can be ruled as a straight line when the heights and distances are drawn as if the Earth's radius was $\frac{4}{3}$ of the true figure (Fig. 5.6).

The diagram of Figure 5.6 shows that for a radio signal from two points which are 120 km apart, if one point is about 650 metres above sea level and the other at least 220 metres high the path can be regarded as just in line of sight. This may be described as 'the grazing ray' even though the geometric line of sight as calculated from the normal Earth radius is close to 200 metres below sea level.

Refraction of radio waves is significantly affected by weather conditions. The effects range from occasions when the signal is actually bent away from the Earth, thus making the radio horizon closer than the geometric horizon, to occasions when refraction is so pronounced that signals can follow the curve of the Earth thus making the radio horizon effectively at an infinite distance. A similar effect applies to visible light but it is so small that it is not usually noticed except when a mirage can be seen.

These conditions of extended radio horizon are most common in the last few hours before a thunderstorm. If you could see by radio waves as we do by light, the Earth's surface at these times would appear as an immense bowl with

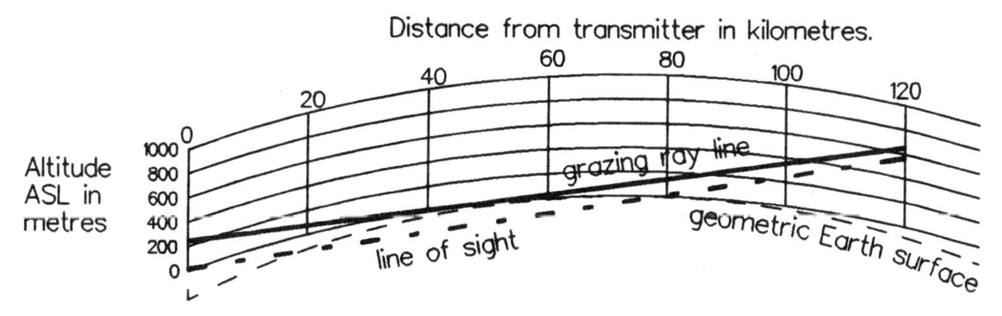

Fig. 5.6 $\frac{4}{3}$ *Earth radius diagram*

your observing point at the bottom of the bowl no matter where you happened to be, and the edges of the bowl would be the edge of the weather cell which in some directions may be up to 500 kilometres away.

The $\frac{4}{3}$ radius Earth model is about right for conditions when the atmosphere is well mixed such as applies on a windy night when there is some showery rain.

There may also be times when refraction effects cause some signals to be trapped within a layer of atmosphere and ducted to wherever the layer goes. This may be a good thing if the ducting happens to connect transmitter and receiver, but the result is more often a temporary loss of signal. For commercial services, ducting is usually a disadvantage, but for amateur operators, there are times when known ducting conditions can be used to make occasional contacts over record-breaking distances.

Refraction also affects the direction from which signals appear to come. This is particularly relevant to ground-wave signals crossing a coastline or mountain range at an angle and is one of the reasons why resolution of direction-finding measurements of MF broadcasting stations cannot be made finer than about 5 degrees. Plotting of ray lines in such cases is exactly the same as for light with the glass block and pins but on a much larger scale.

5.6 REFLECTION

Light reflects from shiny objects. We are aware of this from two sorts of observations:

1. If we look at the shiny object we see an image, which may be a very good likeness if the object is made to be a mirror, or in other cases may be hazy and distorted.
2. Sometimes we see the light as a bright spot which has some of the qualities of an image. A watch face may show a round bright spot in a partly darkened room; a sheet of flat galvanised iron laid out in sunlight may show a rectangular patch on the inside of a shed roof.

The same mechanism that creates reflections of light from shiny surfaces occurs to radio signals but on a much larger scale. The side of a house is usually not big enough to be noticed; a multistorey building or large industrial complex may give effects in the local area up to perhaps a couple of kilometres away; and a cliff face that is a fair proportion of the side of a mountain may give workable signals for UHF communication into a valley up to the range of its visible horizon.

Roughness of surfaces matters less than for visible light; an irregularity must be at least as big as half the wavelength before its effect will be noticed at all. Thus for transmissions in the VHF range where wavelength is in the range of 1 to 10 metres, a hillside which looks to the eye visibly rocky and quite broken may seem to the radio signal as a bright shiny reflector.

This form of reflection can give signals beyond line-of-sight range in mountainous areas for voice communication services. For television, reflections of this type cause ghosting if some of the direct signal is also being received. Occasionally they can be useful if the direct signal can be totally rejected and the aerial aimed at the source of a single clear reflection. For broadband data services, reflections are usually a disadvantage because they give another signal with a time delay which can interfere with the clear definition of logic states.

Total internal reflection can occur for radio signals, but about the only common instance of it is in the case of signals in the HF band being 'reflected' by the ionosphere (see Section 5.8).

There is a form of reflection of radio signals which has no significant parallel in optics. A resonant length of conductor will absorb all the power which comes within its capture area and reradiate it. This form of reflection has a large number of effects—some good, some bad. Resonance occurs when the electrical length of the conductor is close to half the wavelength or an odd multiple of half-wavelengths. The degree of this effect depends on the direction of the conductor, being at a maximum when the conductor is side-on to the incident wavefront, and zero when the conductor is end-on.

Resonant lengths of conductor are used to give directional effects to antennas of the Yagi type. These extra conductors, generally called *parasitic elements*, are cut to exact lengths that are slightly different from the true resonant length to control directivity. An element cut slightly longer will reflect the signal away from its direction; one slightly shorter will lead the signal to greater concentration in its direction.

Resonant reflections are occurring all the time in a host of places but most can be ignored. It is only those that are within a couple of wavelengths of one of the aerial systems and positioned within the directional beam that normally have any practical effect on signals.

The effect of resonant reflections is probably most often noticed in relation to motor vehicles. Every metal piece of the vehicle body and every conductor in the electrical system is potentially a resonant length for some signal. In practice, most of them do not matter; it is only those that match the particular frequency you are presently using that are of concern. When you are in a moving vehicle you will find a host of metal objects causing reflections which impose localised standing-wave patterns on signals. For instance when you are receiving an FM broadcasting station close to the edge of its service area you may notice a regular 'woof, woof, woof' of background noise as you drive; this is the effect of standing waves from localised reflections.

If you particularly want to hear a segment of the program in a place where this effect makes the signal unreadable as you drive, you can stop and drive a few centimetres at a time until the signal comes clear, then stay there for the time of the program segment you are interested in.

For these resonant reflections, the normal rules of reflection (reflected ray at the same angle as incident ray etc.) do not in general apply. The power of the signal is being absorbed and then reradiated, so the 'reflector' is a new point source of radiation.

The reflected signal has the form of a sphere centred on the resonant object. Figure 5.7 shows how standing waves are formed by that effect. This drawing may need some explanation because it is an attempt to show in two dimensions something that is essentially a three-dimensional phenomenon. On the left-hand side of the diagram, the section shown as 'incident wave' shows a side-on view of what the signal would be like if the reflecting object did not exist; a field-strength meter reading anywhere in such a signal would show a constant reading and the meter needle would not move if the meter itself is moved. This section shows a steady flow of radiated signal from left to right.

With the reflecting object in place, a field strength measurement at any point on the lines marked 'reinforcement' would show a steady reading which is higher

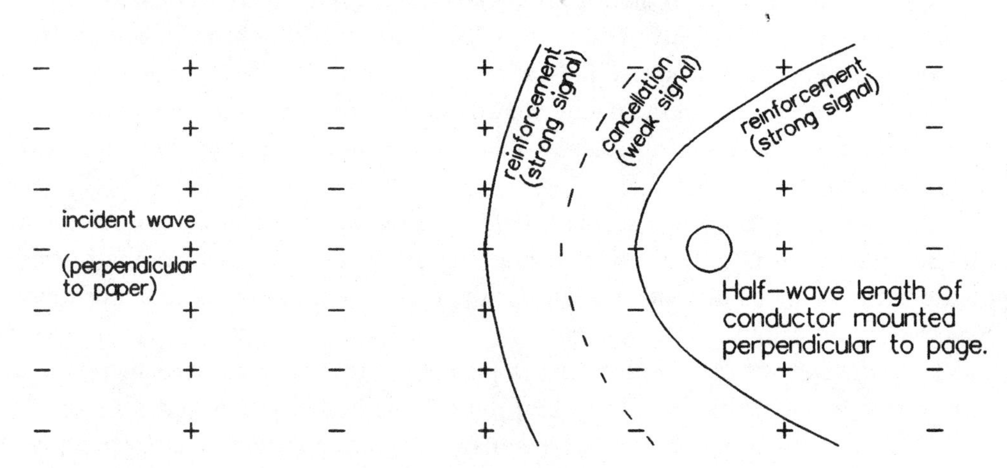

Fig. 5.7 *Mechanism of 'reflection' from a single resonant object*

(theoretically by 6 dB) than the steady reading of the incident wave. A reading on the line marked 'cancellation' shows a similar steady reading but this time significantly lower (by 15 to 40 dB) than the reading for the incident wave. If the meter is moved anywhere along the line of reinforcement, the meter needle will show the same higher reading; and similarly anywhere on the 'cancellation' line, it will show a steady reading at the weak signal level. When the meter is moved either towards or away from the reflecting object, the needle will fluctuate in a wavelike pattern; at any point it is constant over time but changes with changed position.

At a high-power transmitting station, there are many bits of metal that could form resonant reflectors but their effect will depend on the power and wavelength of the signal being transmitted. For a 10 kilowatt MF broadcasting station, for instance, where the wavelength is a couple of hundred metres, the possible effects may be seen out to about the boundary fence of the aerial paddock or a little beyond. For a station broadcasting with similar power in the VHF range, the radiating structure is usually at the top of a high tower and resonance effects are limited to the metalwork at the top of the tower. For most people this will remove the effects from practical consideration, but if you happen to be a linesperson or rigger whose job is to work at the top of the tower, it would be well to remember that stray resonances can cause localised spots of personal hazard radiation at distances up to several wavelengths from the antenna element that is supplying the power.

The most significant effect of reradiated signals around broadcasting stations is the possibility that the structure may include a dirty or corroded joint which will radiate harmonic signals. Joint problems are avoided by designing the station with deliberate bonding between metal parts and bonding to a common earth mat. Usually by the time sufficient bonding has been designed in to prevent harmonic generation, the other effects of resonances are not a problem, although some may cause a few percentage points' change to the directional properties of the aerial system.

5.7 SCATTERING

There are some opportunities for over-the-horizon radio systems based on reflection from minute dust particles, trails of meteorites and air mass boundaries in the upper atmosphere. The mechanism is similar to that which makes a searchlight beam visible in dusty or foggy air.

Scattering may be either forward scatter or back scatter. The forward scatter mode is sometimes used commercially in the form of *tropospheric scatter* links where an isolated community requires several channels of telephony and or data services. In principle, each transmitter illuminates a volume of the atmosphere and each receiver 'looks at' the same volume of atmosphere. A minute proportion of the transmitted signal is slightly deflected somewhere in that volume and is picked up by the receiver.

Path loss for tropospheric scatter systems is much higher than for microwave links operating under radio-line-of-sight conditions. To get a signal of commercial reliability, very powerful transmitters and highly directional aerial systems are needed. The principle is useful only for fixed links.

For commercial use, systems giving a broadband bearer equivalent to 60 telephone channels can be used over distances up to about 200 to 300 kilometres with capability to extend that distance by adding a second or third hop.

For military use, experiments have been done at distances up to 1000 kilometres for one hop. The attraction for the military is that there are no intervening sites where equipment may be subject to sabotage or fall into enemy hands.

Back-scattered signals are generally not used for communication but the mechanism is one of the processes by which weather radar systems collect their information.

5.8 RADIO SIGNALS IN THE UPPER ATMOSPHERE

The outer layers of the Earth's atmosphere take the full force of the Sun's rays, and in the process they absorb most of the ionising part of the radiation. The effect of that absorption of energy is to create and maintain layers of electrically charged (ionised) air in the upper atmosphere. Meteorologists and upper atmosphere scientists identify parts of the atmosphere with letters of the alphabet. The 'A' layer is the bit we walk around in. They can identify what may at some times be up to four separate charged portions in three of those layers.

At the top of the atmosphere there is the 'F' layer, which in the daytime may contain two separate charged portions which are designated 'F1' and 'F2', with F2 at the higher altitude. The F2 layer contains the highest concentration of charge seen anywhere in the atmosphere but also the lowest concentration of air particles, so in times of high solar activity there may be up to one in ten of the air particles carrying a charge. At the altitude of maximum charge, about 300 km, there may be 1 000 000 free electrons (and corresponding charged atoms of air) per cubic centimetre.

The lowest layer that shows any charge is the 'D' layer, and statistics for that are as follows:

1. Height range 60 to 80 km.
2. Free electrons about 100 per cc.
3. Air particles about 100 million million per cc.

The charge on the D layer disappears within a few minutes of sundown each day and reappears again shortly after sunrise. Because of the high concentration of neutral air taking energy from the charged particles, the D layer is an absorber of energy of radio waves but never a reflector. By contrast, the F2 layer absorbs very little and, for some frequencies in the HF and MF bands, is a very effective reflector; its effect usually lasts all through the night.

Between the D and the F2 layers there are 'E' and 'F1' layers which sometimes may reflect or sometimes may absorb radio signals. Signals travelling up from the Earth's surface strike the D layer first. If they are completely absorbed here, the higher layers can have no effect on them. Those which pass the D layer may be affected by the E layer or the F1 layer. Any which pass through the F1 layer may be reflected by the F2 layer and must run the gauntlet of the other three on the journey back to Earth. Any signals that reach the F2 layer and are not reflected by it will pass through and follow the rules for free-space propagation.

Signals which are reflected back to Earth by the upper atmosphere are called *sky-wave signals*. In certain non-technical circles, these are called *skip signals* or *the skip*, but these terms should be used with caution because the word 'skip' is also used to describe a different but related process identifying a region where signals do *not* reach. In conversation you should generally be able to resolve ambiguity by checking whether the speaker is talking about 'skip signals' (the sky wave) or 'skip zones' (where signals are not heard but go overhead).

Sky waves are possible at any frequency between about 500 kilohertz and 150 megahertz. The range of frequencies for which sky waves are economically important, however, is narrower than that—from approximately 2 to 35 MHz. There are some reflections of frequencies higher than 35 MHz but they are too unreliable to be predicted. The order of reliability ranges from a reasonable expectation some time during the day for frequencies around 50 MHz to rare freak at 150 MHz.

The electric charge on each of these layers is produced by radiation from the Sun, so at any time, the concentration of charged particles in each layer depends on the flow of solar radiation at present and in the recent past.

For each two points on the Earth's surface, there is almost always a frequency in the HF band that will give strong clear radio signals between them. Picking that ideal frequency is the secret of effective communication. The catch is that the ideal frequency depends on the interaction of about half a dozen factors, some of which are cyclic and some of which are only partly predictable. These factors are:

1. *Distance*: The ideal frequency rises when stations move further apart.
2. *Time of day*: Higher frequency in the daytime.
3. *Season of the year*: Highest in summer.
4. *Location of reflection point*: Highest when it is directly under the path of the Sun over the Earth's surface; lowest at about latitude 60° either side of the equator.
5. *Rotation of the Sun on its axis*: There is a roughly repetitive factor of conditions on a 27-day cycle.
6. *Internal conditions in the Sun*: There is a long term cyclic effect which can be related to the monthly average of sunspot numbers.

To gain a full understanding of these cycles and the mechanism by which they work would be one of the more difficult learning programs of the radiophysics discipline. It requires study of both meteorology and astronomy with a bit of atomic chemistry thrown in. Detailed prediction of the propagation of each signal requires assessment of the effect of each layer at each point where the relevant signal will strike it. This can be quite a complex calculation.

In each country that uses HF radio systems for over-the-horizon ranges, there is a branch of the government public service whose job is to observe and keep records of the Sun's radiation and to make the prediction calculations. In Australia this body is the IPS Radio and Space Services, which is a branch of the Commonwealth Government. The IPS is a good starting point for all other than superficial inquiries into the long-range use of radio links. For most practical users of HF radio, we are content to leave the details to the specialists at the IPS or one of its associated institutions and use their predictions with just a broad outline understanding of the process and a few rules of thumb.

Within the workable range for any particular set of conditions, there is higher absorption of the energy of the signal for the lower frequencies. The D and E layers do the absorbing, so the signal must get through those before it is reflected. This means that the strongest signals are found at the top (high frequency) of the workable range. At the high-frequency end, there is a *maximum usable frequency* (MUF) for each combination of transmitter and receiver locations. If you were able to establish contact then let the frequency gradually drift higher, you would find that the strongest and clearest signals of all are at the frequency just below the MUF, but that as the frequency drifts higher, you lose contact suddenly and completely. If you were able to quickly move further away from the transmitter or gain altitude in an aeroplane you would find the signal again, strong and clear as before.

The signal of slightly higher frequency is skipping over your head and coming back to Earth further away. The region of country where you cannot hear a signal because you are too close to the transmitter is called the *skip zone*. If you find you are in the skip zone for a particular frequency, you should be able to get good clear signals by choosing the next lowest allocated frequency.

Figure 5.8 shows the paths of relevant signals.

Some of the rules of thumb that are worth following for successful use of an HF communication link are as follows:

1. Experience of conditions in the previous few days gives a good indication of present conditions—especially if both stations are at fixed locations.
2. If the wanted station is heard but only weakly, try the next higher allocated frequency.
3. If the wanted station is not heard at all but the channel appears to be carrying signals from further away, try the next lower allocated frequency.
4. The higher the Sun, the higher the frequency. That is, in the morning when the Sun is climbing towards the zenith, if the signal fades out change to a higher frequency; in the afternoon, if the signal disappears, try a lower frequency.

The ionised layers have an effect on space communications to artificial satellites. Any frequencies that may be reflected or absorbed by any of the layers cannot be used reliably for communications to the other side of these layers. This implies that space communication frequencies are all in the VHF and higher frequency ranges.

Even for frequencies well up into the UHF range, the ionised layers have an effect causing refraction similar to the way a stick appears to bend when it is placed in water. This can make the apparent position of a satellite in the sky to be shifted, and for use with a highly directional aerial system, the shift can be more than the beam width of the aerial. This, of course, must be known and allowed for at the time of setting up the communication link.

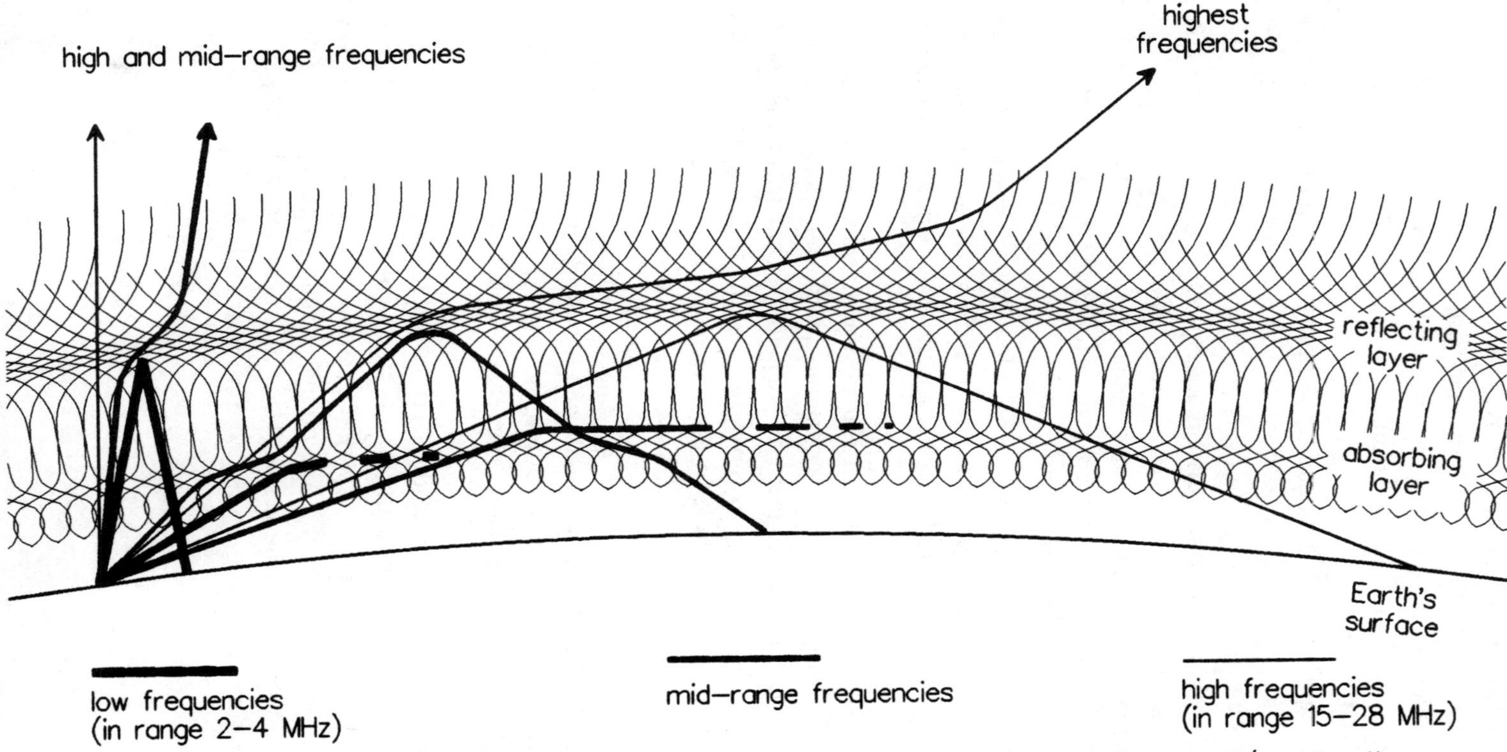

low frequencies (in range 2–4 MHz)

mid–range frequencies

high frequencies (in range 15–28 MHz)

In this diagram lines of different width indicate the path of signals of different frequency/wavelength. Thickest lines indicate longest wavelength (lowest frequency).
The distance across the Earth's surface represented by this diagram is about 2000 km. At that scale the height of the highest mountains is comparable with the width of the line for 'mid–range' frequencies.

Fig. 5.8(a) *Diagram showing the effect of distance from the transmitter on HF radio signals*

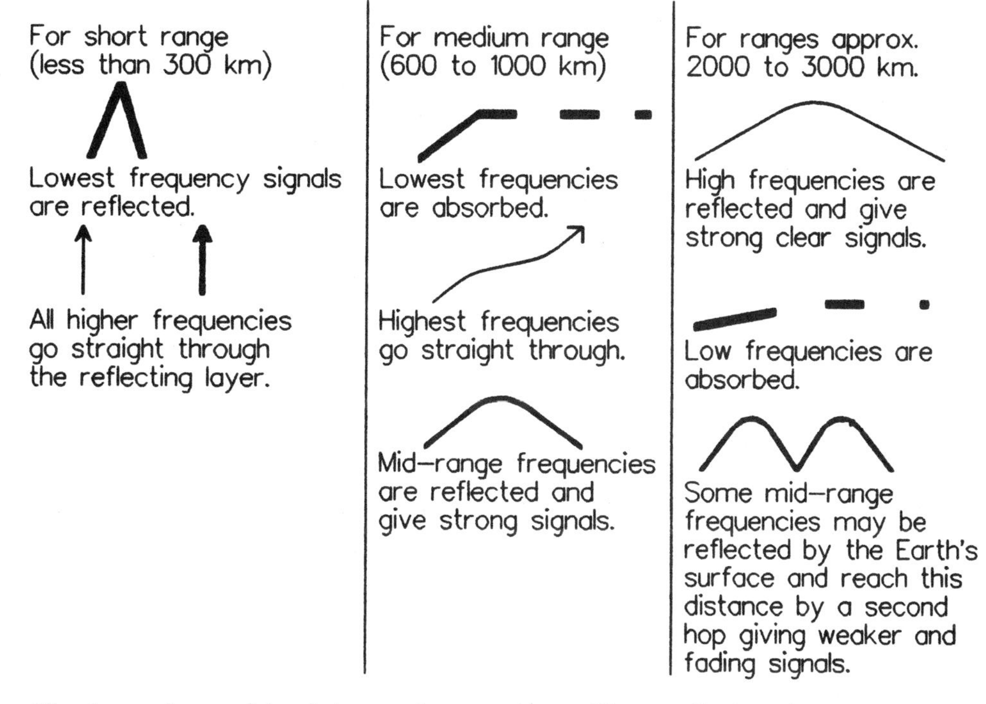

The longest possible distance for one hop with practical aerial systems is about 3200 km. Beyond that range all signals propagate by reflection from the Earth's surface and a second or more hop.

Fig. 5.8(b)

CHAPTER 6

HOW THE BANDS ARE USED

This chapter is written assuming you have read and understood Chapters 2 and 5.

6.1 BRIEF OUTLINE OF THIS CHAPTER

The listing detailed in this chapter will give some information on the mechanisms most important for propagation in each band. There is also some outline data on the practical uses of each band. The spectrum is considered in order from the lowest frequencies to the highest.

6.2 VERY LOW FREQUENCY (VLF) BAND

Frequency range: 30 kilohertz and below
Wavelength range: 10 kilometres and longer

Ground wave is the major factor in propagation for the VLF band. During all daylight hours, the sky wave is completely absorbed by the D layer. The effects of diffraction and refraction are not separately discernible, but to the extent that they do occur, they will increase the strength of ground-wave signals. A reflection would require an object at least half a wavelength (minimum 5 km) in its longer dimension to be placed an appreciable portion (at least $\frac{1}{10}$) of a wavelength above ground level.

Use of this band requires enormous aerial structures, and even then they are electrically short. This means that high losses due to ground resistance must be accepted, and high-power transmitters must be used to overcome these losses. The size of aerials and power required means this band is unsuitable for mobile transmitters. Narrow bandwidth dictates that it is used only for information of very low data rate. Once launched, however, the ground wave will carry a stable, reliable signal many thousands of kilometres.

Time signals and navigational beacons are ideal uses. The very lowest frequency signals are the ones that penetrate soil and water the best. Some frequencies just above the audio range are used for communication with submarines under water. The band is also used for communication to mines and tunnels by setting very large coils to couple via the induction field.

6.3 LOW-FREQUENCY (LF) BAND

Frequency range: 30 to 300 kilohertz
Wavelength range: 10 kilometres to 1 kilometre

As for the VLF band, ground wave is the main mechanism of propagation for the LF band. The range of ground-wave signals for a particular degree of loss is reduced as frequency rises, but with sufficiently powerful transmitters and large aerial structures, a stable signal over several thousand kilometres is possible. At 300 kilohertz, an unloaded quarter-wave whip is 250 metres high, so for some frequencies the largest practical structures can be made to give reasonably efficient radiation. Directional arrays are not generally possible. Aerials are too large for efficient use by mobile services except in the case of shipping where the vehicle is often large enough to accommodate the length required. There are also some aeronautical services at frequencies close to 300 kHz; for these the inefficiency of aerials is accepted and the expected range is scaled down as required.

At the low-frequency end of this band, aerial efficiency rises as higher frequencies are chosen which tends to offset the reducing range of ground-wave signals. So for an installation in which there are practical limitations on aerial size and transmitter power, the range achieved may be almost independent of frequency.

Bandwidth restrictions are not as severe as for VLF but still lead to the major use of this band being for telegraphy services—either hand-keyed as for CW radiogram, or machine-generated as in the case of beacon transmitters, remote weather stations etc. In some parts of the world there have been a few voice channels made available for long-range broadcasting. The early experiments with voice broadcasting used frequencies in the LF range.

6.4 MEDIUM-FREQUENCY (MF) BAND

Frequency range: 300 kilohertz to 3 megahertz
Wavelength range: 1 kilometre to 100 metres

This band has effectively been split into three sections. In practically all countries of the world there is a large section in the middle of the band devoted to AM broadcasting. The low-frequency limit of this allocation is usually between 520 and 530 kHz and the other end somewhere about 1500 to 1600 kHz. The section from 300 to 530 kHz contains the international marine radiotelegraphy allocations and a host of radiolocation beacon transmitters. Aeronautical non-directional beacons are usually allocated a spot in this range. There are at least three or four of these associated with each commercial airport and each requires a channel clear of other transmissions over a several hundred kilometre radius for effective use by aircraft direction-finding equipment. The section from 1.6 to 3 megahertz is often used in conjunction with the low-frequency end of the HF band for reliable short to moderate range radiotelephony using single sideband transceivers.

For frequencies below 1.6 MHz in the daytime, ground wave/diffraction is the only propagation mechanism of interest, but over this range its effectiveness decreases as frequency rises. At 1.6 MHz, a 1 kilowatt AM signal radiated from a quarter-wave vertical aerial would be so weak as to be unreadable at less than 200 kilometres distance for most receivers over most types of soil.

Where AM broadcasting is required over a wide area of rural or pastoral countryside, the designer will seek allocation of the lowest available frequency and aim for maximum allowable power. On the other hand, a station designed for local coverage of, for instance, a country town or regional city would be allocated a frequency near 1.6 MHz and just sufficient power for the coverage need.

At night-time, the sky wave does begin to have an effect mainly due to reflection from the E layer, but in the AM broadcasting section of the band, the effect is no help for propagation in the intended service area.

Imagine that a receiver is placed initially close to the transmitter and then moved to progressively greater distances and the signal strength monitored. At the starting point, the signal via ground wave is much stronger than the sky wave so the received signal is strong and steady. With each move outwards, the ground wave is weaker. Eventually a location is reached where both are about the same strength. The catch is that at this location, the two signals do not necessarily add strength to each other. There will be some places where they arrive out of phase and cancel—this will cause fade-outs of signal in particular locations. What is probably more annoying for committed listeners is that the layers of charged air drift and change in a fashion similar to visible clouds of water vapour, and a receiver permanently located anywhere in this zone of about equal signal strengths will be subject to severe fading from time to time. At greater distances, the sky wave may be clearly stronger and there may be a zone where steady signals can be received at night-time but nothing at all in the daytime.

This limitation is not avoided by increasing transmitter power; it is the ratio between ground wave and sky wave that causes the effect. Special aerial structures have been developed for transmitter sites to concentrate radiation in the horizontal plane and limit the strength of the sky wave. This has the effect of placing the first zone of cancellation at a greater distance than for a non-directional aerial. In a well-designed system, an *antifading radiator* of this type may approximately double the radius of the primary service area.

The MF band is used in the mobile mode but normally only in association with fixed stations. Noise due to distant thunderstorms and thermal atmospheric effects is quite high so a relatively high transmitted field strength is needed to overcome it. At the receiver, however, a supersensitive system is no advantage, so quite small whip antennas will give all the signal that can be used. Thus a fixed transmitter with a highly efficient radiator teams well with a mobile receiver with a rudimentary signal-collection device. There is some mobile-to-mobile operation in the sense of ship-to-ship channels towards the 3 MHz end of the band. Ships are about the only vehicles physically big enough to support efficient aerial systems for wavelengths in this range.

6.5 HIGH-FREQUENCY (HF) BAND

Frequency range: 3 to 30 MHz
Wavelength range: 100 to 10 metres

There are intense demands on the high-frequency band because it is the one that offers long-range possibilities using sky-wave propagation.

In the range of frequencies between 1.6 and 35 MHz, there is a gradation of character of the reflected signals. At 2 MHz, signals are highly reliable but weakened due to absorption in the D and E layers. In times of high sunspots, the useful range for signals at 2 MHz in the daytime is about 150 km. This extends to about 500 km at night-time and perhaps a bit more in times of low sunspot numbers. There is no skip zone for 2 MHz signals and they are almost never stopped by propagation disturbances although at times signals may be weak and noisy.

At the other end of the HF band in the range between 18 and 28 MHz, there are frequencies which give strong clear signals worldwide but not for all of the time. They can, however, be teamed with a lower frequency to give reliable coverage on the basis that the propagation conditions which make one unusable will enhance the performance of the other. Therefore, in the range between 2 and 28 MHz there is a trade-off between highly reliable but weak short-range signals at 2 MHz and strong clear long-range but less reliable signals at 28 MHz.

Each of the cyclic factors listed in Chapter 5 (time of day, season, Sun's rotation or sunspot cycle) may be capable of changing the ideal frequency by a factor of about 2:1, and they have a multiplying effect on each other. Thus, for any given pair of points, the ideal frequency in conditions of summer daytime at the peak of the sunspot cycle may be 6 or 8 times as high as would be ideal for a winter's night at the time of sunspot minimum. At any one time, the range of frequencies either side of the ideal that is workable covers about a 2:1 or 3:1 ratio for short (less than 500 km) ranges, or down to 1.5:1 for a longer range.

Communication services using the HF band commonly are allocated several frequency channels over a wide range of frequencies. Note the following examples:

1. A basic business communication service from a base to a small group of mobiles operating at about 500 km range may be allocated two frequencies: one in the 3 to 4 MHz range and another about double that. The higher frequency would be used for daytime communication, the lower at night.
2. A similar business with operations ranging from 0 to 2000 km distance may require three frequencies with the lowest somewhere between 2 and 3 MHz, and the highest close to 12 MHz.
3. A short-wave broadcasting service aiming to give reliable daytime coverage for a particular region between 3000 and 3500 km from the transmitter may have frequency allocations with the lowest close to 9 MHz and the highest at 26 MHz, and include two or three other frequencies spaced within that range.

4. An emergency service offering safety-of-life coverage may be allocated up to six frequencies spread over the range from 2 to 25 MHz, and monitor all of them 24 hours per day.

If you are starting a new service, for instance a base and mobiles SSB transceiver system, you will have been asked at the time of licence application for the locations of the stations. Frequencies allocated will have been chosen to suit average conditions over the known range of the sunspot cycle for those locations. The question is: what is the best channel for first calling in conditions that apply now? If you refer your allocated frequencies and locations to the prediction service, they can give you a chart that relates the most useful frequency to time of day for the average conditions of the current month. For initial tests, their suggestion is the frequency you should use.

Once you establish contact or if you can prove that a signal transmitted to a working receiver was not heard, then a couple of simple rules of thumb will help you get the clearest signals:

1. If you can hear the signal you want but it is only weak, try the next higher frequency.
2. If the signal you want is not heard at all but you can hear signals on the channel that appear to be coming from further away (for instance you can hear a foreign language), try a lower frequency.

The aim of frequency selection is to get as close as possible to the MUF but not so close that contact is made unreliable by minor drifts in conditions. Figure 6.1 shows a graph of signal strength versus distance.

There are many services making demands on the HF band: mobile transceiver networks; short-wave broadcasting; amateur, maritime, aeronautical, radioteletype and picturegram links; and many others. There are incompatibilities between them but all require access to frequencies over a wide range. Each of the many different types of services is allocated several narrow portions spread through

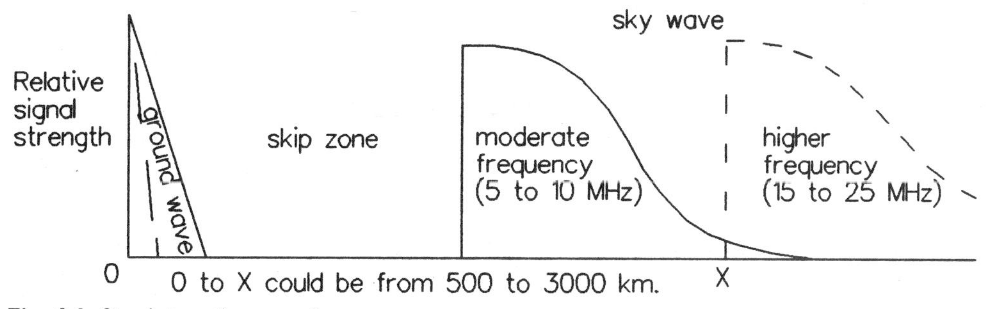

Fig. 6.1 *Signal strength versus distance*

the HF spectrum. Some of these allocations are less than 100 kilohertz wide with provision for only a dozen or so channels. Because of these requirements, the HF band more than any other is characterised by a multitude of narrow allocations.

Once a range of frequencies is allocated to a particular service, the decision of which to use at any particular time is usually made by the operators on the spot. In some cases, automatic machinery is used to poll all frequencies allocated and decide on the strongest and clearest channel; this process is called *automatic link establishment*.

6.6 VERY HIGH FREQUENCY (VHF) BAND

Frequency range: 30 to 300 MHz
Wavelength range: 10 metres to 1 metre

For the VHF band, the major mechanism by which signals can go further than the visible horizon is refraction in the atmospheric layer immediately above the Earth's surface.

During times of peak sunspot activity, the low-frequency end of the band is within the usable range for sky-wave propagation over very long paths. There are always significant skip zones and the mechanism is not generally reliable enough for commercial use. The amateur fraternity and those people interested in identifying and recording long-distance reception of television (TV Dx clubs) make use of sky-wave propagation in the VHF band.

In general, ground-wave propagation is not significant although it can in theory be shown to occur. Its practical effect is probably to add 2 or 3 kilometres to the usable range of each transmission. There are some cases where television signals can be received a few hundred metres past the brow of a hill or mountain range and it is observed that the strongest signal is very close (1 to 3 metres) to the local ground surface. This results when a strong clear signal reaches the brow of the hill and completes the last few hundred metres of its journey by ground-wave propagation.

Reflections do occur but only in the case of very large objects such as cliff faces on a mountain side. Reflections are more prevalent at the high-frequency end of the band.

The VHF band is mainly used for short-range reliable signals over flat or slightly rolling terrain with little or no gaps in coverage. The effective range is generally reckoned to be 'visible line of sight plus a little bit'. The 'little bit' is significantly larger for the low-frequency end of the band than for the high-frequency end. For conditions that would give a 300 MHz signal a range of 50 km, a signal of the same power and aerial gain at 30 MHz may have a range up to 150 km.

At the edge of the usable range, there is a significant fringe area where signals of lower reliability are available. In the 160 MHz range of frequencies, for instance, there is a band allocation for mobile communications working to a fixed base (taxis, delivery companies etc.) in which the spectrum managers seek to guarantee interference-free use for a 40 km radius. Business uses of this band notice disruption if the grade of service goes below about the F90,90 level. This means that a workable signal is available in 90% of locations 90% of the time. To achieve this grade of service over a 40 km radius requires that the base aerial be located on a reasonably prominent position.

At the other end of the reliability scale in the 2 m amateur band immediately adjacent in the spectrum, there are particular operators who have had quite long series of contacts over 300 to 500 km distances by picking the most appropriate time of the day and accepting the fact that on some days contact will not be possible. Rare contacts at ranges of 2000 km and above are reported by amateurs interested in breaking records. There is thus a trade-off between distance and reliability over the range of distances from about 20 to 2000 km.

Since the advent of widespread use of computers, electrical noise has been a factor to be considered in the VHF band. The noise which comes from harmonics of all the square-wave signals in the computer, and particularly the monitor, is prevalent in all densely populated areas and especially in the central business districts of major cities. This noise spreads through the whole VHF band but is worst at the low-frequency end. There is also noise associated with high-voltage power distribution lines which is most noticeable in the corridors between power stations and large industrial centres.

Width of the band in terms of frequencies available is less of a limitation than for the lower frequencies. Some broadband services such as television are possible but there is still considerable market pressure for access to channels. The bulk of the spectrum space of the VHF band is devoted to two main uses:

1. *Television*, which for each channel requires several megahertz of bandwidth. It is a high-volume user of the band worldwide to the extent that each major city has at least three or four VHF TV stations and most provincial centres are covered by one or two.
2. *Single-channel voice communication services*, most commonly using a fixed base in contact with a group of mobiles over a radius of up to 40 km. These services are the second biggest user of spectrum space in this band.

These two major users between them take up about 75% of the band. Other significant but smaller allocations are for aeronautical and maritime local communications, broadcasting of high-fidelity audio programs, and uses by amateur operators. There are also small allocations for radionavigation uses and, towards the high-frequency end, for space uses and radioastronomy.

In the 1990s, uses of the VHF band still show the effects of history and marketing fashion which probably means that in the long term there will be some changes in demands and allocations before the most economic uses of the band are settled on. Until the time of the Second World War (1939 to 1945), the VHF band was largely unused although a lot of it was allocated. Technology made low frequencies usable first so the band has tended to fill from the bottom end (of frequencies).

Administrators have so far not given due recognition to the larger coverage area possible with low-frequency allocations, so they charge the same licence fees for high frequencies as for low. A gradation of licence fees to reflect the more restricted coverage and lower demands on spectrum space of higher frequencies will give opportunity for those users who need to cover only a small area to make economic gains by clearing the channels with large-area capability. In particular, low-power local uses of television in the UHF band will at least partly replace high-power stations in the VHF band which give regional coverage but really only market their services to the city of their location.

6.7 ULTRA-HIGH-FREQUENCY (UHF) BAND

Frequency range: 300 to 3000 megahertz
Wavelength range: 1 metre to 10 centimetres

As with the MF band, the UHF band also has major subdivisions within it. For many countries there is an allocation for UHF television giving about 70 channels or so somewhere in the range between 500 and 900 MHz. The frequency range below the UHF TV allocation carries the same sort of traffic as the top end of the VHF band and can usually be treated as an extension of it. Frequencies higher than the TV allocation generally carry broadband multichannel services such as *demand assigned multiple access* mobile services, or *point-to-point multiplexed telephony links* (microwave bearers).

The $\frac{4}{3}$ radius Earth model is often used with fixed links in the UHF band to define the expected limit of workable range, although for lower reliability and depending on weather conditions, the limit of range can still be two or three times that figure. Propagation at the high-frequency end of the band is significantly more restricted to line of sight than at the low end.

Ducting and reflections have much more effect on the UHF band than they do on the VHF band. Multistorey buildings are large enough to reflect workable signals over a couple of kilometres, so on city streets, the signal usually arrives at street level by reflection. This may have some effect on the bit error rate of high-speed data services, but for data at bit per second rates equivalent to a voice channel, the time delay will usually be too short to notice.

Ground-wave and sky-wave propagation mechanisms have no practical effect on the UHF and higher frequency bands. The upper atmosphere can affect signals passing through for communication with satellites. There is some of the bent stick effect which makes the apparent position of the satellite change. The charged layers affect polarisation of the signals which means that aerial systems must accept random polarisation.

For the UHF band and all higher frequencies, the limit of sensitivity of the receiver is set by thermal noise generated internally by the first amplifying stage of the receiver electronics.

Dipole antenna elements are physically small so multi-element systems with high directivity and gain are normal for fixed links. For mobile use, highly directional systems are normally unusable (except in the case of steered arrays) so gain is limited to what can be obtained with two or three collinear elements. Physically small aerials have limited capture area so mobile systems tend to be near the low-frequency end of the UHF band and designed for short to moderate range (close to line-of-sight conditions).

There are some uses for mobile systems in the 1500 MHz frequency range particularly related to mobile communication with artificial satellites. For frequencies above that range, use is almost entirely by fixed point-to-point links with highly directional aerial systems.

Usage of the UHF band is even more affected by recent history than is the VHF band. In the mid-1990s, the development of new more effective components and circuit techniques is still actively in progress and the workability limits of a whole range of practices are being steadily pushed up in frequency. This probably means that final definition of the most efficient uses of the UHF and higher bands is still some time away.

6.8 THE MICROWAVES

Frequency range: higher than 1 gigahertz
Wavelength range: shorter than 300 mm

Although some over-the-horizon mechanisms of propagation are known, most such as forward scatter are only rarely used. Most use of this range of frequencies is for radio-line-of-sight fixed broadband links with transmitter power less than 1 watt and very high-gain (typically 40 decibels or more) highly directional aerial systems.

For terrestrial systems these are usually either multiplexed voice channels or time division multiplexed data. The power per channel or per bit of data is microscopic, being engineered for a maximum free space path loss equivalent to about 40 to 50 km.

The $\frac{4}{3}$ radius Earth model is commonly used to define the radio-line-of-sight condition. The small wavelength and high towers normally used mean that the signal is usually clear of ground effects so free-space conditions give a close estimate of signal strength.

Atmospheric effects must be allowed for. Ducting is very common for these frequencies and its usual effect on fixed links is to cause fading. A margin of signal strength of between 25 and 40 decibels is usually needed to give the required reliability.

If the ends of the link must be further apart than is possible with a single link, a repeater can be used at a point which is in radio-line-of-sight with both points. If an even longer distance is required, more than one repeater can be used and chains up to several thousand kilometres long can be created with multiple repeaters, each of which is in radio-line-of-sight contact with the one either side in the chain.

With the advent of communications satellites, the technology of broadband microwave links has been adapted for use with satellites in geostationary orbits. In this case, the free space path must be at least 36 000 kilometres, and intermediate repeaters are not possible. Gaining the extra range requires higher transmitter power, much larger aerials and a lower bandwidth to concentrate more power into each channel or data bit. Sometimes receivers must be cooled with liquid nitrogen to keep internal noise low.

There are quite a large number of spot frequencies spread through the microwave bands which have significance for other than communication purposes. Microwave ovens need a frequency set aside for them—they nominally use 2.45 gigahertz (GHz). All microwave ovens can share the same frequency but the generator section is rudimentary compared with a communications transmitter, and frequency control is poor, so a band around 2.45 GHz must be set aside for their use. Some industrial uses of microwaves are slightly dependent on frequency so there are several other narrow bands besides the 2.45 GHz allocation for similar applications.

In the same way as the spectrum of sunlight has dark lines called *Fraunhofer lines* which are the signatures of particular chemical elements, there are particular frequencies in the radio spectrum which are the signatures of chemicals. In the microwave range there are frequencies related to compounds as well as to elements, so there is a host of spot frequencies possibly involved. A few of these, in particular frequencies that signal compounds of hydrogen and carbon, are of great importance to radioastronomers and other scientists. One famous example is the hydrogen line at 1.421 GHz.

These lines are only spot frequencies with no sidebands at all; however, a band must be allowed around them because of Doppler effects. If a star is moving at 30 kilometres per second, which is not an unusual velocity for a star, the signal from its hydrogen will be shifted in frequency by 142 kHz. Each of the lines that are set aside for use by radioastronomers must be at least half a megahertz or so either side of the 'rest' frequency to cater for Doppler shifts.

There is a broad range of frequencies above 22 GHz which are absorbed by water vapour in the atmosphere. These frequencies are no good for any long-range use, but for coverage of an area up to the size of a large factory they can be used to advantage. The absorption ensures that interference from more distant systems will not be a problem, and the wide bandwidth available is of interest to the data networking industry. Wireless LANs (local area networks) commonly use frequencies whose range is limited by atmospheric absorption.

CHAPTER 7

RADIATING STRUCTURES, AERIALS, AND ANTENNAS

This chapter is written assuming you have read and understood Chapter 3, and Section 1.6, 'Frequency', Section 2.2, 'Wavelength' and Section 2.5, 'Sine waves'.

7.1 DIAGRAMS OF SOME TYPICAL RADIATORS

Figure 7.1 shows some diagrams of some typical radiating structures.

7.2 INITIAL CLASSIFICATION

The detailed design of conductor arrangements for efficient radio transmission and reception is an enormous subject and this chapter will only outline the principles. There is a book published by the American Radio Relay League called *The ARRL Antenna Book.* It is A4 size and about 30 mm thick, and is crammed with designs but still does not have most of the designs which are used by international short-wave broadcasting transmitters and commercial telecommunications carriers.

Hopefully after you have read this chapter you will be able to look at a particular installation and have some knowledge of the principles of its design; be able to make a reasonable estimate of what it may be used for and its approximate operating frequency; and in the case of directional arrays recognise the likely direction of maximum signal.

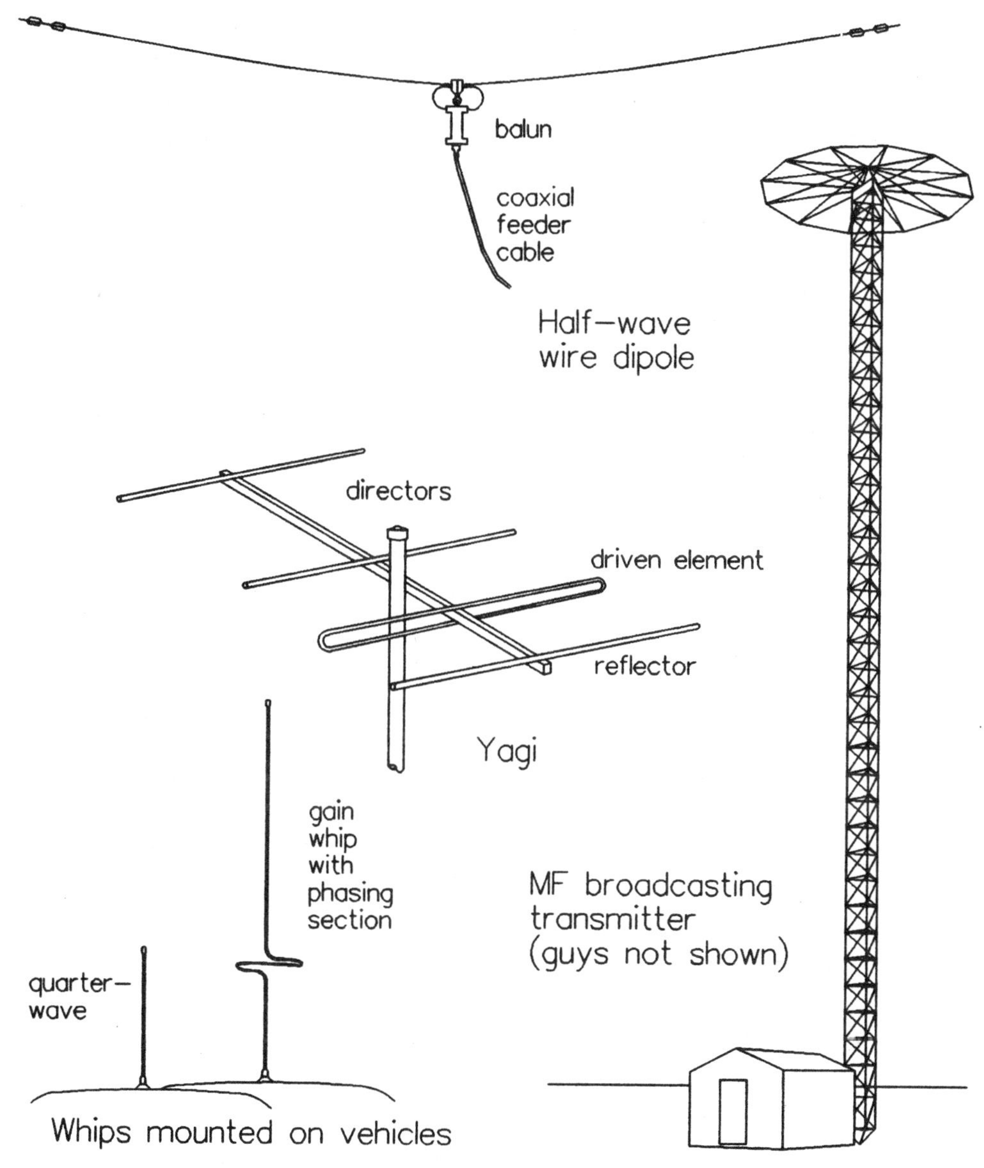

Fig. 7.1 *Some typical radiating structures*

A radio aerial, whether for transmitting or receiving, is a transducer which is the general name for devices used to translate energy from one medium to another. A transmitting aerial translates the energy of alternating currents and voltages in the medium of an electrical conductor into waves of magnetic and electric stress in free space (or vice versa for a receiving aerial).

Every piece of electrically conductive material has some degree of coupling with radiation fields in free space. There is some translation of energy in the transmit direction every time an alternating current flows in a conductor, and some degree of reception whenever a conductor is placed in a radiation field. However, it is only when the conductor is specifically arranged to provide a high efficiency of coupling between mediums that we would normally describe it as a radio 'aerial' or 'antenna'. The general aim in the design of radiating structures is to place conductors in positions so that they can couple freely to the open field then arrange electric currents to flow in them so that resistance losses are small compared with the total power flow.

There are many thousands of individual designs of efficient radio aerials with a wide range of principles of operation. To make the understanding of it all a bit easier, there are some classifications possible so that designs with a similar operating principle can be grouped together.

One of the most basic divisions is between those designs which are 'resonant' and therefore tend to be tuned to a particular operating frequency, compared with those which are 'non-resonant' and so will work over a wide frequency range. Another important division can be made between those systems designed for all-round coverage compared with those which are intended to have directivity. Combining these two factors enables you to sort all the possible designs into one of four groups:

1. resonant, omnidirectional
2. resonant, with directivity
3. non-resonant, omnidirectional
4. non-resonant, with directivity

7.3 RESONANCE

If you have a long narrow bowl partly filled with water, you can place your hand approximately in the centre of the bowl and use it as a paddle to move the water back and forth. If you move your hand very quickly you will do lots of work and splash some water around close to your hand but the main body of water will not move much at all. If you move your hand very slowly, the water will simply flow from side to side of your hand and the main body of it will stay still.

Now if you gradually increase the rate (frequency) from that slow movement you will find one particular rate when the whole body of water moves in unison—your hand moves easily with little relative movement between your hand and

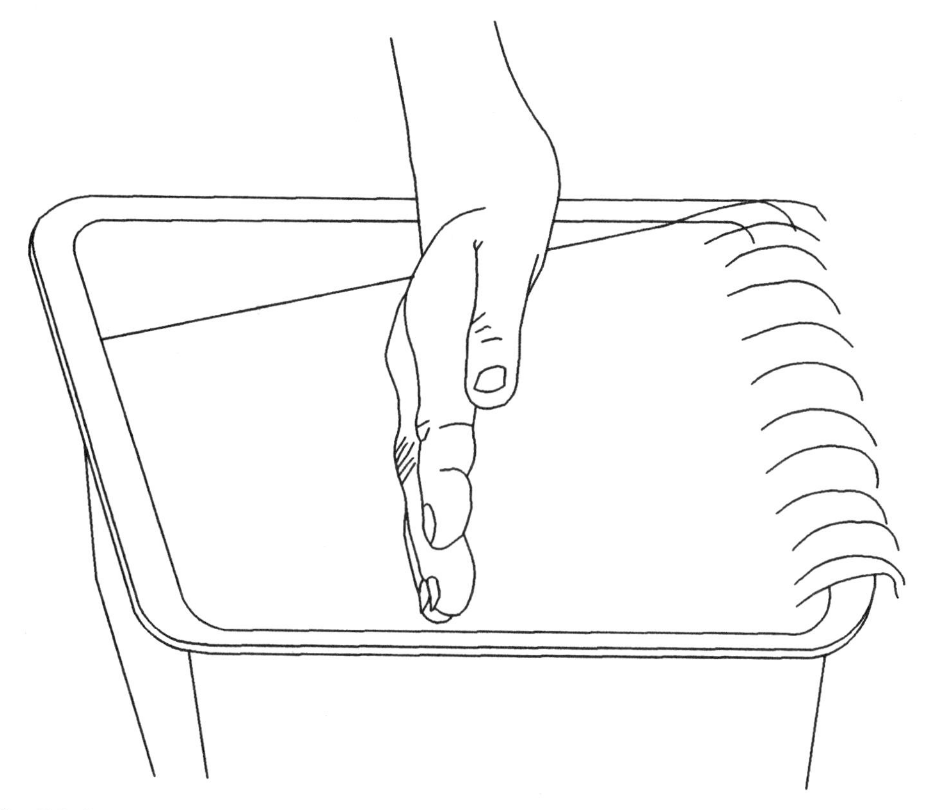

Fig. 7.2 *Demonstrating the principle of resonance*

the water, and you can without much effort build up a large enough surge to slosh over the ends of the bowl (Fig. 7.2).

Imagine now that a half-wave dipole aerial is a 'bowl' of electrons and your hand/paddle is the central coupling point. The same sort of resonance effect takes place.

7.4 THE DRIVEN ELEMENT

The most basic of all resonant radiators is the half-wave dipole. A straight piece of conductor very slightly less than half the length of one wave of the frequency being transmitted or received is just the right length for current to flow back and forth in step with the wave. There is also a cyclic voltage which appears at the ends of the conductor resulting from the surge of electrons piling up as the current flows towards each end. Figure 7.3 shows the voltages and currents in a half-wave of conductor.

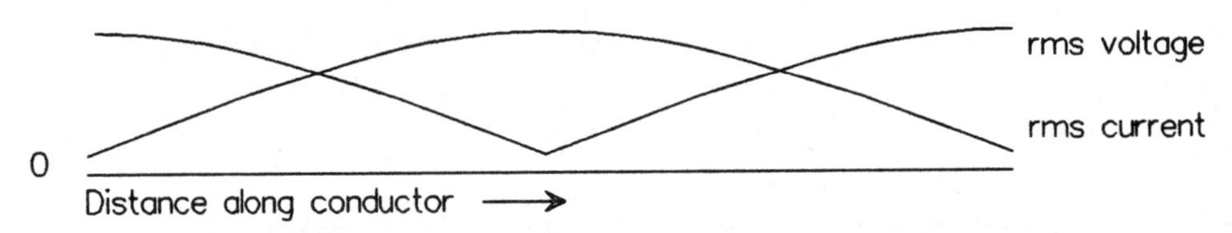

Fig. 7.3 *Voltages and currents in a half-wave of conductor*

At first sight the suggestion that the same piece of conductor could have different currents flowing at different points along its length sounds ridiculous. However, it is not a steady (direct) current; electrons are surging back and forth in the wire in a very similar way to the sloshing of the water described in the previous section. Differences of current are accounted for by the slight elasticity of the electron gas.

A very good illustration of what is happening can be seen by watching full-woolled sheep packing into a race with the gate at the end shut. Each sheep's wool is a bit springy and each will individually move to equalise pressures from all the other sheep. Each time the dog barks there is a compression wave that travels along the mob and packs each one in a bit tighter. When the dog is quiet, all sheep relax and drift back to a less crowded position. The one at the blind end does not move to any great extent but the one nearest the dog moves backwards and forwards several steps.

To couple electrical power to or from this radiating element, connections must be made that match the current and voltage conditions at the point of connection. If the connection is made by breaking the conductor in its middle and connecting a wire to each half (see Fig. 7.4), the current is high and the voltage low (Fig. 7.3) so this matches to a fairly low impedance. For a conductor of zero resistance in free space with no connection to anything else, the half-wave dipole with this form of connection at the centre behaves as if it were a pure resistance of 78 ohms. This is a true resistance in that it dissipates power—not as heat but as radiation. It is called *radiation resistance*.

If the connection is made instead to the end of the half-wave conductor, the voltage is maximum and the current minimum. A dipole connected in this way can present a radiation resistance to the external circuit of up to 3000 ohms. By making connections at other points along the length of the conductor, the

Note: measurements described as 'rms voltage' or 'rms current'. The specification 'rms' is derived from electrical technology and indicates an alternating signal that has the same heating effect (i.e., flow of real power) as a DC signal giving the same meter reading. Its significance in relation to Figure 7.3 is that it is the indication that would show as a steady reading on the face of an appropriate meter if the line was tested at that point.

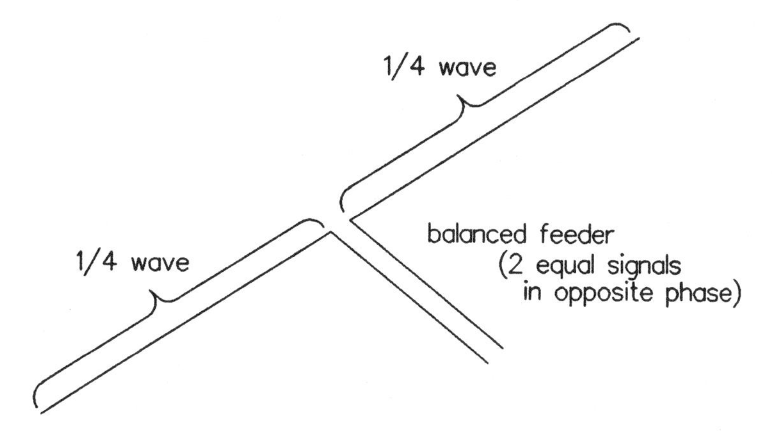

Fig. 7.4 *Centre-fed half-wave dipole*

impedance can be made to appear as any figure between 78 and 3000 ohms. Other objects close to a half-wave dipole will change the radiation resistance, generally reducing its ohmic value. Most items have only a slight effect but another resonant piece of conductor close to the dipole and parallel to it can reduce radiation resistance to 20 to 30 ohms.

The dipole does not radiate or receive equally in all directions but its radiation pattern can be easily and accurately related to an isotropic radiator. Radiation and sensitivity are at a maximum in all directions at right angles to the direction of the conductor, and at a minimum end-on. In theory there is a point of zero radiation exactly end-on, but in practice, it is so small that no one ever finds it.

If you search for this null point in a well-designed test situation, you may be able to identify a point that is 30 to 40 dB down compared with the direction of maximum response. In the rough and tumble of the real world, reflected signals from nearby objects and other factors may limit the depth of this null to about 15 dB.

Figure 7.5 shows diagrams called *polar diagrams*. They are graphs of field strength for a typical half-wave dipole. Note that to get a full understanding of the pattern you need to visualise the combination of these graphs in three dimensions; the difficulty is that we do not yet have three-dimensional pieces of paper to draw on.

The field strength distribution graphed here is often called a *doughnut* pattern because in three dimensions, it is the shape of a slightly flattened doughnut with the half-wave of conductor inserted through the centre of the hole.

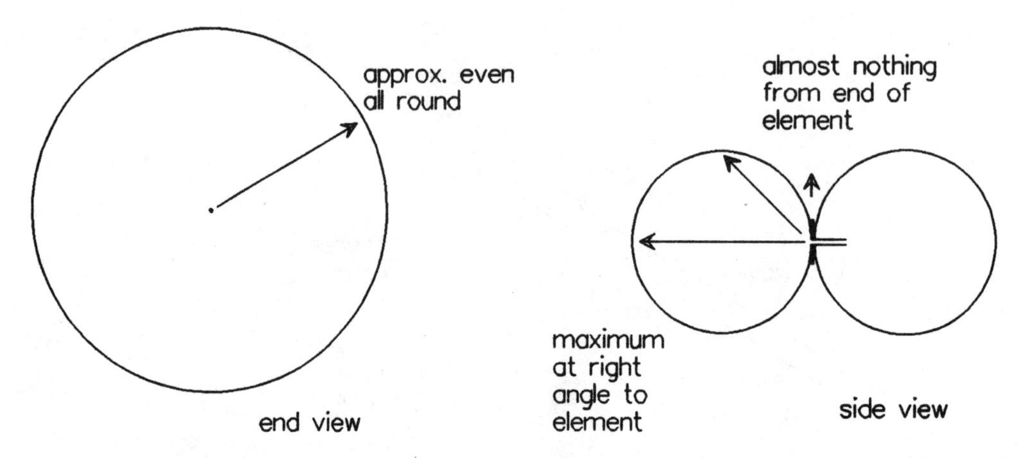

Fig. 7.5 *Polar diagrams of field strength for a typical half-wave dipole in free space*

7.5 UNDERSTANDING POLAR DIAGRAMS

A polar diagram is a type of graph. In a normal graph with x and y coordinates, the values of each coordinate are represented by straight lines which are at right angles to each other. In a polar diagram, one of the coordinates is represented by a series of concentric circles and the other by straight lines radiating out from the common centre. A polar diagram can be related to a normal graph by thinking of the x-axis as the central point with the various values of x being the radial lines. The concentric circles represent the values of y. (See Fig. 7.6.)

When polar diagrams are used to display directivity of aerial arrays such as shown in Figures 7.5, 7.10 and 8.10, the radial lines represent relative direction

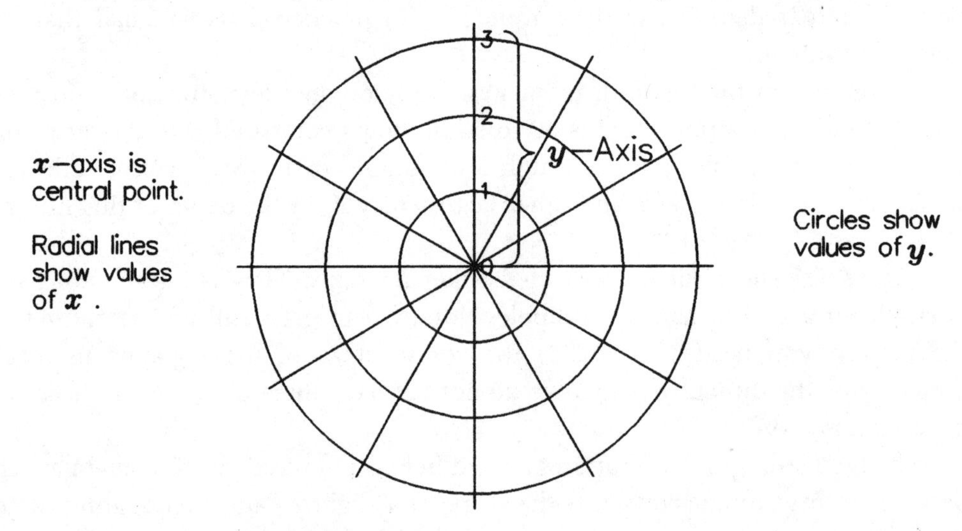

Fig. 7.6 *Polar diagram showing coordinate system*

and the concentric circles represent relative field strength. It is very tempting to look at a polar diagram and think of it as an illustration of how far the signal will go in each direction. That is *not* its intention, although of course in a lot of cases there will be a tendency for both factors to have similar-shaped curves.

Directional effects take place in three dimensions and the polar diagram only displays conditions on one plane. For full specification of a particular aerial pattern, at least two diagrams are needed: a 'horizontal' and a 'vertical'.

7.6 POLARISATION

The current induced in a conductor by a radiation field is in a particular direction dictated by the magnetic and electric field components of the radiated signal. If that direction is across the conductor rather than along it, very little if any current will be induced. For line-of-sight conditions, the signal is generally maximum if the driven element of the receiver is parallel to the driven element of the transmitter. (See Fig. 7.7).

This aspect of the signal is called *polarisation*. For terrestrial signals, polarisation is generally either vertical or horizontal. An attempt to receive a vertically polarised signal with a horizontal receiving aerial, or vice versa, may give a loss of sensitivity of up to 30 dB.

When signals are reflected or refracted, the polarisation may be changed, not necessarily always in a predictable way. It is well to remember that if the signal

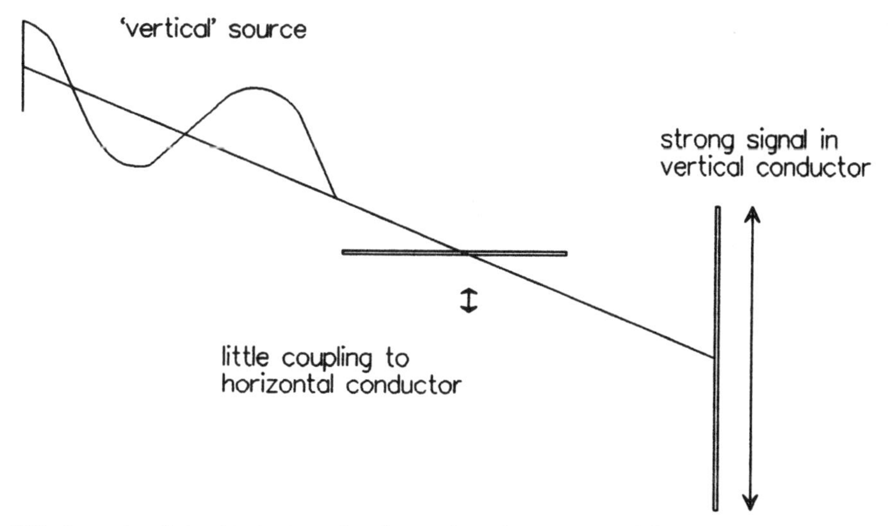

Fig. 7.7 *Strengths of signal in horizontal and vertical conductors received from 'vertical' source*

strength does not come up to expectations, then polarisation is one of the factors that needs to be checked.

For satellite communications, polarisation is liable to be in any direction, so systems must be made sensitive to all possibilities. Systems of crossed dipole elements are often used. There are also antenna systems that can launch the signal so that the field vectors rotate as the radiated wave proceeds. This method of operation is called *circular polarisation*. A circular polarised signal can be received equally well by either horizontally or vertically polarised dipoles with a power loss of 3 dB. Circular polarisation can be either right-handed or left-handed rotation. An attempt to receive a right-hand polarised signal with a left-hand polarised antenna will show exactly the same cross-polarisation effect as trying to receive a vertical signal with a horizontal dipole. Circular polarisation overcomes the effects of rotation that happen when signals are reflected or refracted in passing through ionised layers of the atmosphere.

Polarisation of radio waves has the same mechanism as polarisation of light, but the application of the effect shows up in different ways for the different frequencies. The mechanism of polarisation and the theory behind it is extensively covered by any study of the science of optics. A radio signal is a coherent wave and the effect of polarisation on radio is closely allied to the same effects for coherent light. The majority of the observed differences between polarisation of radio signals and polarisation of light are due to the incoherent nature of ordinary light.

7.7 ANTENNA GAIN AND DIRECTIVITY

For practical aerials, there will be some directions of the hollow sphere (referring to the model mentioned in Section 3.4) in which the signal strength is reduced, and if the power is not absorbed and lost as heat, then it must go in other directions to make that part of the signal stronger. That effect can be deliberately used to make the power output of the transmitter more effective in the useful direction.

Extra elements can be added to simple resonant antennas to work as either reflectors or directors to reduce the radiation in directions where it does no good, and concentrate the beam towards the receiver. For non-resonant structures, concentration of signals in wanted directions is usually increased by making the physical dimensions larger (that is, assuming the thing was pointing in the right direction in the first place!).

Figure 7.8 shows how directional effects are formed.

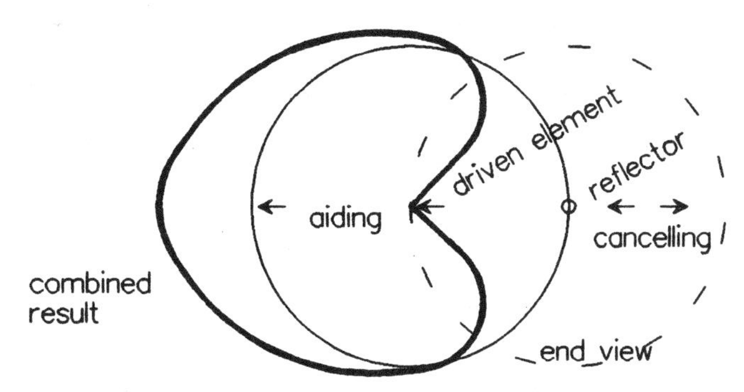

Fig. 7.8 *How directional effects are formed*

For most wire or rod antennas it is fairly easy to concentrate the power in the wanted direction by 10 times, and difficult but often not impossible to achieve a 50 to 100 times increase. This concentration of the power is called *antenna gain* or *aerial gain*, and is measured in decibels. A gain of 10 times corresponds to 10 dB, a 50 times gain corresponds to 17 dB, and a 100 times gain corresponds to 20 dB. There are some large horn or parabolic dish radiators which have up to 40 dB of antenna gain equal to concentration of the power by 10 000 times.

Transmitter installations are often specified as having a certain *effective radiated power*. This is calculated by multiplying the electrical power of the transmitter by the aerial gain (with losses in feeders etc. subtracted).

In some cases such as a point-to-point link where all power radiated in other than one particular direction is lost, aerial gain may be all that is needed for a complete specification of performance. In other cases where radiation is needed in several different directions, the gain figure may need the addition of directivity information such as a polar diagram. An example may be a vertical element used either as a mobile whip or a broadcasting transmitter aerial where the signal can be concentrated in the vertical plane but must give all-round coverage of the horizontal.

The effect of concentration of the beam is easy to demonstrate and not too difficult to understand in the case of the radiation from a transmitting aerial but may be less obvious for reception. In practice, however, all the additions that result in gain in the transmitting direction produce the same gain for reception. There is a principle called the *reciprocity theorem* which states that any function of a transducer which applies to changes in one direction can be equally applied in the reverse direction. In this aspect, the reciprocity theorem ensures that antenna gain figures calculated for transmission can be equally applied to reception.

7.8 PHASE

For alternating voltages or currents, the relative timing of the waveforms is important. Compare the two diagrams in Figure 7.9.

If the two waveforms in each of the diagrams represent, for instance, the voltages on two adjacent tracks of a piece of printed circuit, in one case the voltage difference between them is zero but in the other it goes as high as the peak of both waves added together. This aspect of phase has obvious importance for the insulation requirements in electrical circuits.

Phase relationship of the two components is the factor that generates the beat patterns illustrated in Section 2.7 and causes the multipath fading described in Section 5.3.

In relation to radiated signals, phase affects the result of adding two signals; the result at a receiver can be addition which makes the received total 6 dB stronger than either signal alone or it can be complete cancellation. Directional properties of aerials and antennas are almost all the result of signals from several directions being in phase in the wanted direction and out of phase in all other directions.

7.9 PARASITIC ELEMENTS

A resonant length of conductor placed close (within half a wavelength) to a driven element will have a marked effect on the resulting signal. The extra element will absorb power and reradiate a signal, which in some directions will

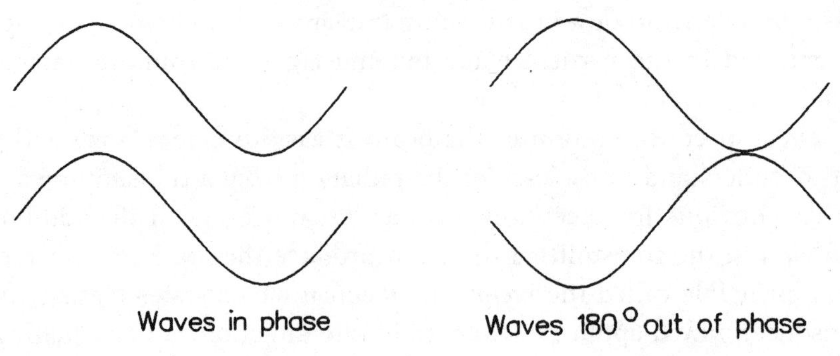

Fig. 7.9 *Waves in phase and 180° out of phase*

be in phase with the original and add to it, and in other directions will be out of phase and cancel. The exact effect of a particular conductor depends on the exact length of the conductor and its spacing from the driven element. In general, extra conductors that are slightly longer than exact resonance will reflect signals away from the direction they are in, and if they are slightly shorter than exact resonance they will increase the signal in their own direction. Spacing between elements is usually in the range from 0.1 to 0.3 wavelengths.

All parts of a system such as this that affect the directivity of radiation but are not directly fed with power are called *parasitic elements*. Those that are intended to minimise radiation in their direction are called *reflectors*; those that maximise signal in their direction are called *directors*.

Yagi type antennas use a driven element, a reflector and a number of directors—from one to about 8 or 10—and are capable of giving antenna gain of up to about 12 dB. *Corner reflector* antennas use several reflectors and can give antenna gain of about 8 or 9 dB with no directors at all. For higher gain, the functions of Yagi and corner reflector antennas can be combined.

7.10 MAJOR AND MINOR LOBES

Recall the information in Section 7.4, 'The driven element'. If the half-wave dipole radiator were set to work as described in that section but then had another half-wavelength of wire attached to one end, it would form an aerial system with somewhat changed directional properties. Each of the half-wave sections would produce radiation with a polar distribution as for the half-wave dipole; but currents in the two sections are exactly out of phase with each other, and so from the point of view of a distant receiver exactly at a right-angle to the conductor, the effects of the two sections cancel to give zero signal.

There are other directions in which radiation from the two sections is in phase and therefore aiding, and in those directions signal is actually slightly stronger than the maximum from a half-wave dipole. The polar diagram of the signal from a full-wavelength aerial is similar to that shown in Figure 7.10.

If a half-wave dipole is arranged with an extra half-wave of conductor on each side (so that the combined length totals $1\frac{1}{2}$ wavelengths), then a similar but more complex pattern of lobes and cancellation nulls is produced. In this case, radiation at a right angle consists of two signals from half-wave sections which cancel each other, and one section which is not cancelled. However, the power from the transmitter is now split between the three sections so the field strength in that direction is reduced below that of a half-wave dipole. As in the case

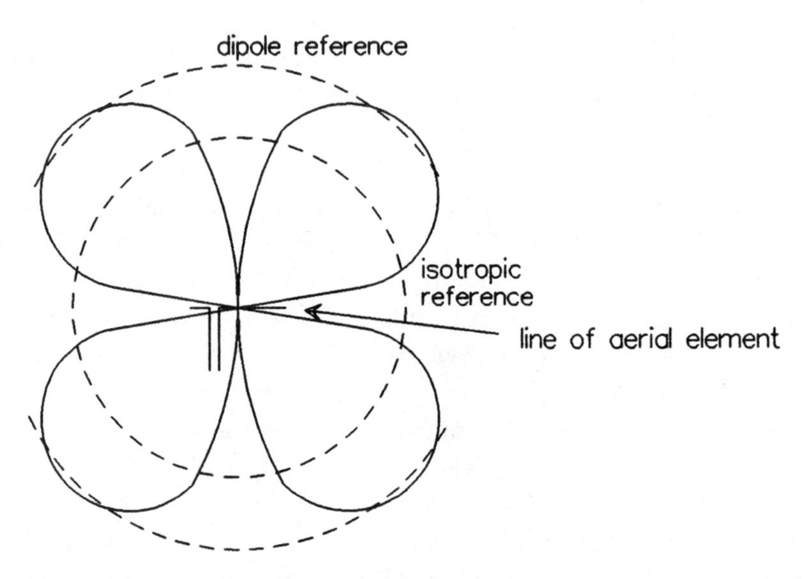

Fig. 7.10 *Polar diagram of the signal from a full-wavelength long wire aerial*

shown in Figure 7.10, there is a direction in which radiation from all three sections is in phase, adding up to a field strength greater than that of a dipole. The direction of this *major lobe* is closer to the line of the conductor than in the case of the full-wave aerial.

In the polar diagram for a $1\frac{1}{2}$-wave aerial, there are nulls on each side of the right-angle between the minor lobe at the right angle line and the major lobes (see Fig. 7.11).

For long wire aerials which are many wavelengths long, there are narrow, concentrated major lobes close to the direction of the wire and a larger number of nulls and minor lobes in other directions. The shape of this range of nulls and minor lobes is sometimes likened to the petals of a sunflower (see Fig. 7.12).

In a variation of the theme of the long wire design, the conductor can be separated into half-wavelength sections, and phasing components added between each section so that currents in all sections are in phase. This forms a *collinear* aerial with a narrow concentrated major lobe all around at right angles to the direction of the conductor and a range of nulls and minor lobes in other directions. This form of construction is commonly used as a vertically polarised aerial for mobile communications bases to give an omnidirectional horizontal response with gain of between 6 and 10 dB.

The nulls and minor lobes pattern is caused by the signal coming from two or more sources which are physically separated by more than a wavelength. The principle is exactly the same as that of an optical interferometer, and, in the

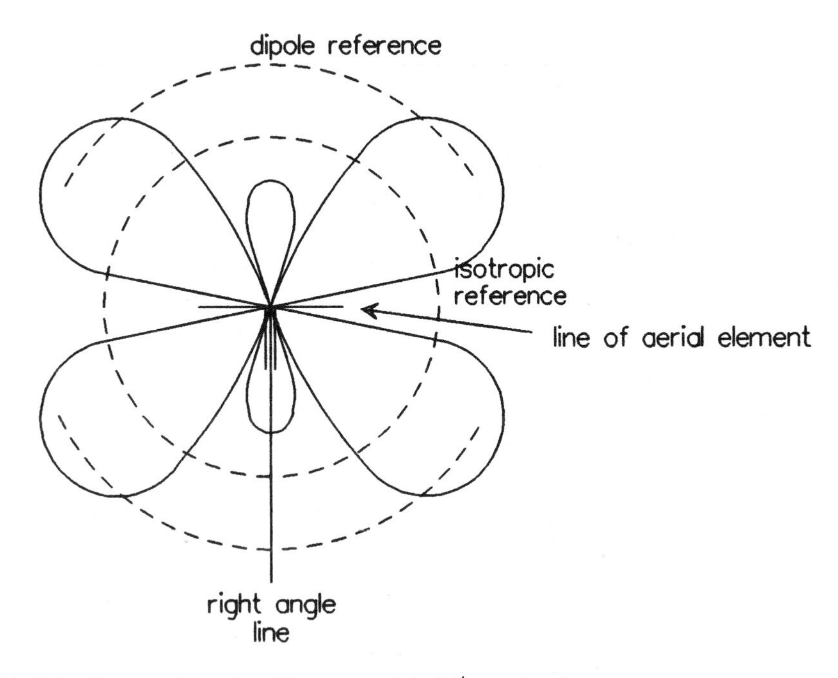

Fig. 7.11 *Polar diagram of the signal from an aerial of $1\frac{1}{2}$ wavelengths*

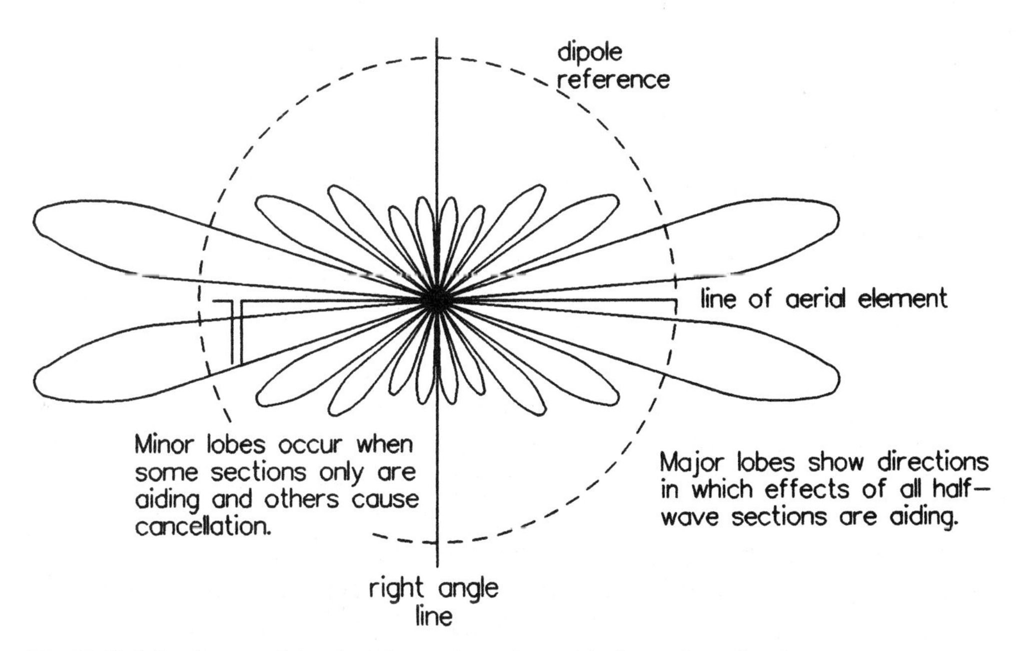

Fig. 7.12 *Polar diagram of the signal from a long wire aerial of several wavelengths*

same way that the optical instrument can be made to measure finer angles by building it with greater separation between collecting points, the pattern of nulls and lobes becomes finer and more frequent as the aerial structure is made physically bigger.

All radiating and receiving systems which are large in terms of wavelength will generate a pattern of minor lobes which in almost all cases are unwanted because they take power from the main beam and send it in unwanted directions. One of the finer points of design is that as much as possible the power in a minor lobe due to one section is cancelled by a minor lobe from another section being launched out of phase.

7.11 FEED LINES

In general you cannot simply connect a piece of wire to the feed point of a driven element and expect it to carry radiofrequency power to the intended place. The piece of wire would itself act as a radiating element coupling signals in both directions. For an aerial system with gain, that would be disastrous. To ensure the feed line (also called the *transmission line*) carries the signal without radiation, it must have conductors arranged in a particular way; the general aim is that there should be at least two conductors running close to each other and parallel and that equal and opposite currents are flowing in the conductors.

The great majority of transmission lines can be classified according to physical construction into one of three types:

1. coaxial cable
2. balanced open wire line
3. waveguide

For operators of small transceivers (such as fixed base-and-mobiles radio-communication networks) in the UHF and lower frequency bands, the feeder is almost certain to be a coaxial cable. Open wire lines are used for very high power installations (over approximately 10 kilowatts) where the high-voltage insulation required is difficult to provide with coaxial cable, and waveguides are used with microwaves where the diameter of a coaxial cable would be a significant proportion of a wavelength (frequencies of 2 GHz and higher).

Although most people are more familiar with coaxial cables, the principles of operation of a transmission line are easier to understand when explained in relation to a balanced open wire line, and so the next two paragraphs refer to that form of construction.

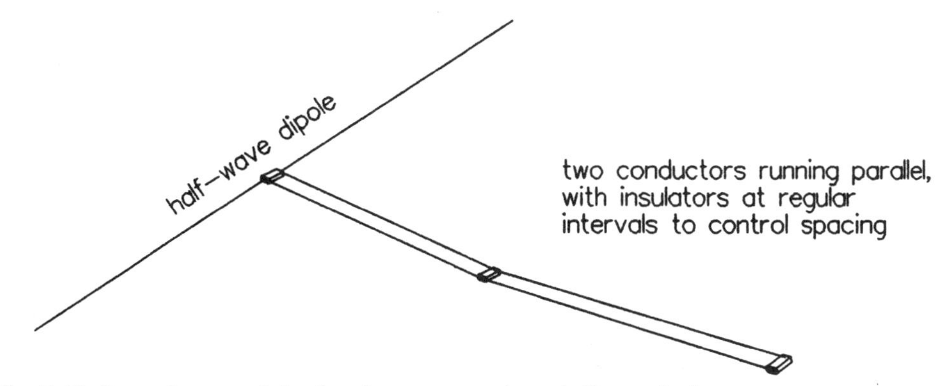

Fig. 7.13 *Open wire transmission line shown connected to a half-wave dipole*

In the case where the driven element is a half-wave dipole, one simple way of feeding radiofrequency power to it is to run two identical wires from the centre point of the dipole to the transmitter output or receiver input and to connect one wire to each side. (See Fig. 7.13.)

In the transmitter, the line must be fed by a centre-tapped transformer or another circuit that has the same function. The two wires of the line carry equal but opposite currents and the voltages on them are also equal and opposite if compared to a reference earth. There is, of course, a tendency for each wire to radiate a signal due to the fact that it is a length of conductor carrying radiofrequency currents and voltages. However, from the point of view of a distant observer (any position more than about half a dozen times the distance separating the two wires), there is a cancellation because in all cases the signal radiated from one wire is exactly matched by an equal and opposite signal from the other. The effect of having two wires carrying equal and opposite currents and voltages is that each prevents radiation from the other and the power of the signal is carried along the line with little or no leakage.

A coaxial cable also has two conductors insulated from each other. One is a wire or rod running through the centre and the other is a hollow tube or braid around it, constructed so that it is a certain distance away from the centre conductor (Fig. 7.14).

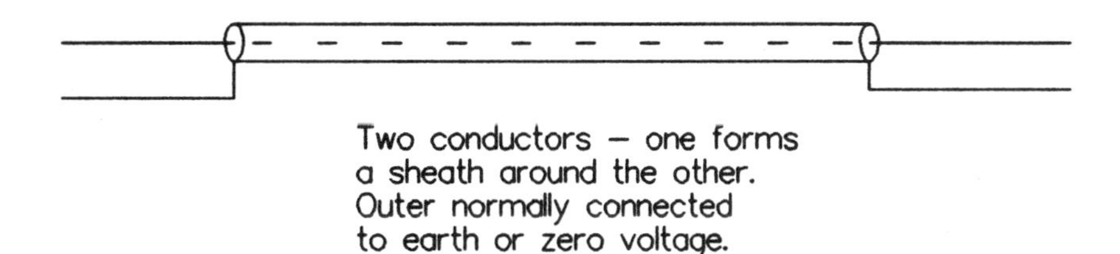

Fig. 7.14 *Schematic diagram of a coaxial cable*

As in the case of the two-wire line, there are equal and opposite currents flowing in the two conductors, so radiation of the magnetic vector is prevented by exactly the same cancellation effect as before. Voltage conditions are a bit different—the outer conductor is normally earthed which has the effect of a Faraday shield around the voltage applied to the inner conductor.

In the form in which the outer tube is a braid of fine wires, coaxial cable can be mass-produced at a cost comparable with that of multiwire cables of similar size, and can be installed almost as simply as a single wire of similar flexibility. Because of this convenience, coaxial cable has become the almost universal means of transferring signals of low to moderate power in the frequency range up to 500 megahertz. Coaxial cables are generally better shielded than open wire lines of equivalent performance; it would not be possible, for instance, to bundle a group of open wire transmission lines together as is routinely done with coaxial cables. In the case of most small transmitters and transceivers where the driven element is a half-wave dipole, coaxial cable is used and the connection is arranged in such a way that one side of the line can be earthed. A *balun* transformer (this means *bal*anced to *un*balanced) can be used (see Fig. 7.15) or there are a number of ways of constructing the driven element itself to accept an unbalanced connection.

Waveguides are tubes (usually made of copper) which may not look all that much different from tubes for carrying air or water. The important factor to note about them is their size in relation to the wavelength of the signal they carry.

Imagine a two-wire line with a quarter-wave section of line connected to it at a right angle. (An explanation of the effect of a quarter-wave transformer is given in Section 7.14.) If the quarter-wave section is short-circuited at the far end, it appears to the two-wire line as an open circuit and the signal being

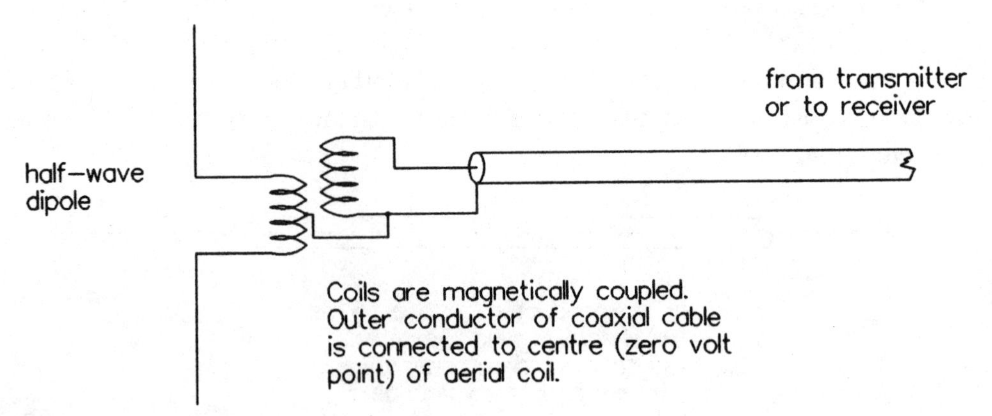

Fig. 7.15 *Method of coupling a coaxial cable to a half-wave dipole*

carried on the line will flow straight past it with no change. There can be more than one of these short-circuited quarter-wave sections connected to the line, in fact there can be lots of them, and none of them will take anything away from the signal because all appear as open circuits. You may get a rough idea of the operation of a waveguide by thinking of it as a balanced two-wire line with a large number of short-circuited quarter-wave sections connected along its length. (See Fig. 7.16.)

Tubes of many different shapes can be used for waveguides. The important point for a waveguide is that the distance across the face of the longest dimension should be at least half a wavelength and less than about three-quarters of a wavelength.

Detailed explanation of how radiofrequency power is propagated along a waveguide soon becomes heavily involved in mathematics and requires some understanding of the basic physics of electric currents. Libraries carry technical books on the subject under the classification number 612.

7.12 STANDING-WAVE RATIO (SWR)

This section is quite technical and involves some mathematics. It is, however, the point about connecting transmitters to aerial systems that most confuses many users. To fully understand it, you will need to have some basic knowledge of electrical principles to the extent of the detail of Section 10.6, 'The water analogy'. If it does not directly concern you at this stage, you may go straight to the next chapter without losing the thread of the story.

Whenever a transmission line or waveguide carries a signal over more than about $\frac{1}{20}$ of a wavelength, it imposes its own conditions on the ratio between voltage

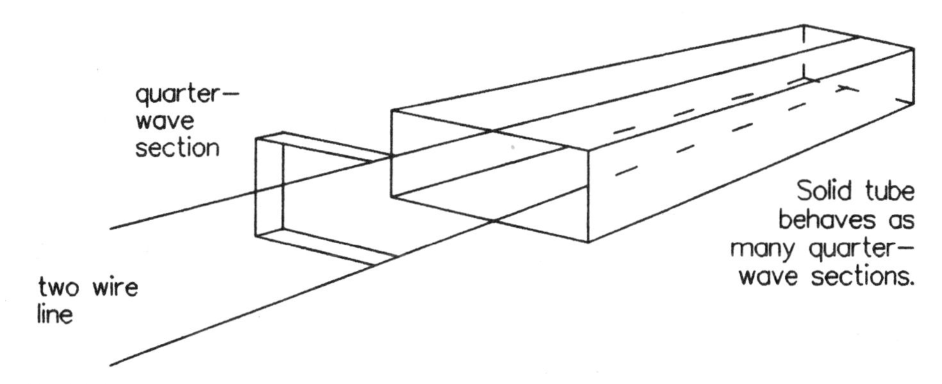

Fig. 7.16 *Waveguide principle of operation*

and current that it will accept. There is series inductance in the lengths of conductor, and capacitance between the conductors, and the effect of these combined means that each piece of transmission line has a *characteristic impedance.* If the transmission line were infinitely long, the application of a certain voltage to its open end would mean that a certain current would flow. To that extent, the line at its connection point behaves as if it were a resistor of a particular ohmic value.

From the point of view of the electronic circuit supplying the power, the effect of an infinite length of transmission line would be exactly the same as a particular value of resistance (power leaves the driving circuit and never returns); however, it is called impedance, not resistance, because the power travels along the line rather than being converted into heat or radiation.

Characteristic impedance of a transmission line is determined by the size and spacing of the conductors and the dielectric properties of the insulation material. For a quite wide range of frequencies, the impedance does not depend on frequency. We have, for instance, coaxial cable which is specified as 50 ohm impedance and will behave as a 50 ohm impedance for all frequencies from less than 100 kilohertz to over 500 megahertz. A cable that looks very similar may be specified as 75 ohm impedance; the difference is that the inner conductor is made of slightly finer wire. Mass-produced transmission lines are commonly described by their characteristic impedance.

When a load is connected to a transmission line, the load itself has a certain impedance. The example of most common interest is when a line carrying power from a transmitter is connected to the driven element of an aerial or antenna. If the driven element is properly tuned, it will present a load equal to its radiation resistance to the end of the line.

In a few fortuitous cases, the radiation resistance will be exactly equal to the characteristic impedance of the line. In those cases, all of the voltage that arrives will appear across the load and all of the current will be absorbed by the load as it arrives. Power flow is all one way. In most cases, however, the radiation resistance does not at first exactly match the impedance of the line. A certain flow of power has already been accepted by the line from the transmitter and it arrives at the aerial on a coming-ready-or-not basis, but it cannot all be accepted by the radiation resistance.

The principle of physics that energy can be neither created nor destroyed applies at this point. The power (energy) is arriving at a certain rate; it cannot all be accepted, so the excess is reflected back along the line to the source. The transmission line of course requires that the reflected power be transmitted with the impedance conditions that it is designed for. The example shown in Figure

7.17 refers to a common type of aerial installation which is described in more detail in Section 8.3.

To show what is happening with actual numbers, take the case in Figure 7.17 where a 50 ohm coaxial cable is terminated in a load that has exactly 35 ohms radiation resistance, and all reactance has been tuned out. Assume that exactly 1 amp of radio frequency current is flowing if the line is exactly matched which of course means that there is 50 volts applied and incident power is 50 watts. When the incident power arrives at the aerial, the 1 amp of current flowing through 35 ohms would produce only 35 volts across the load and dissipate only 35 watts. The actual conditions that apply at the load are that the current flowing rises slightly above 1 amp, the voltage rises above 35 volts in proportion to the increased current, and the power absorbed by the load increases above 35 watts to not quite 50 watts. Power not accounted for flows back down the line towards the source.

The voltage that actually appears across the load in this example is 41.25 volts and current is 1.18 amps which accounts for 48.7 watts of the original 50 watts. The other 1.3 watts flows back down the line as 7.75 volts and 0.18 amps.

Phase relationship between incident and reflected power is important. In some places, voltages are in phase and will add to give a figure higher than that supplied by the transmitter. There are other places where they are out of phase and subtract from each other, reducing the voltage that can be measured at that point. These voltages can be measured with a radiofrequency voltmeter. With a probe that can be moved along the line, the needle of the meter is seen to swing over the range from maximum to minimum voltage in a regular wavelike fashion as the probe is moved (Fig. 7.18).

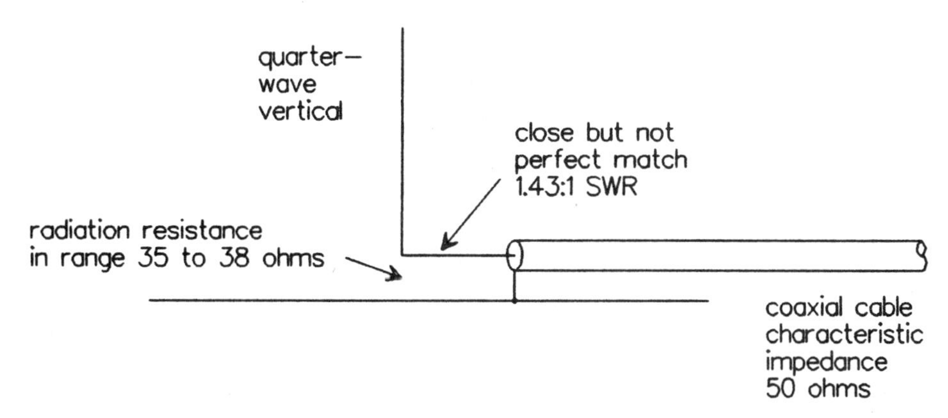

Fig. 7.17 *Line and load with mismatch*

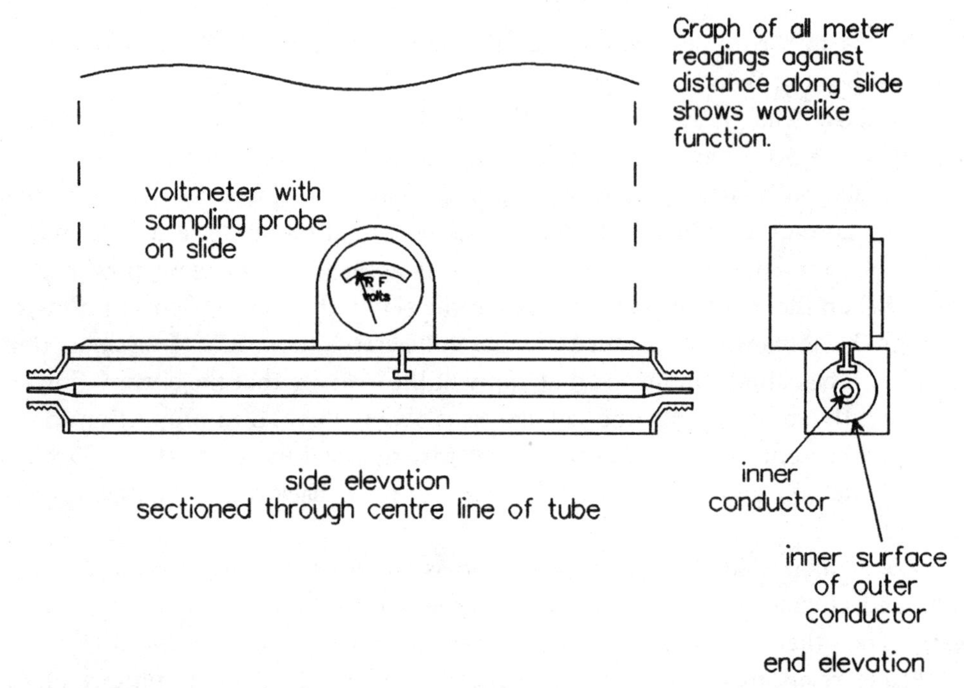

Fig. 7.18 *'Slotted line'—a laboratory instrument for directly displaying and measuring standing waves*

The meter reading at any particular point is steady; the waveform is seen only by comparing readings at a number of different points, for instance by sliding the probe along the slotted line. The waveform of all the meter readings along the line is called a *standing wave*.

If there is a good match between the radiation resistance and the characteristic impedance of the cable, the lowest voltage measured in the standing wave will be almost the same as the highest. That condition is described by saying that the standing wave ratio is close to 1:1. If the match is worse, the ratio between voltages will be bigger to the extent that if there is either a complete short circuit or a complete open circuit at the load, the ratio between maximum and minimum voltages of the standing wave will be from double the voltage supplied by the transmitter to zero. The theoretical effect of a short or open circuit load is to give a standing wave ratio of infinity to 1.

Note that the load resistance can be either higher or lower than the characteristic impedance of the line and may include some reactance (which shows up if the relative phase between current and voltage is measured at that point). The measurement of standing wave ratio gives an indication of the degree of mismatch but no direct indication of which way adjustments have to be moved to correct it. Measurements of the ratio between maximum and minimum voltages of the standing wave give a good indication of whether the antenna is accepting and radiating all of the power that is carried to it by the transmission line. Measurement of SWR is a good proof-of-performance test for a system that is working well, but is less useful as a diagnostic test for fault-finding.

The instrument shown in Figure 7.18 is designed to be connected in a coaxial cable transmission line between the generator of the signal (often a transmitter) and the load (which could be an aerial array). It is an inconvenient instrument to use in the field; directional power meters or meters calibrated to read SWR directly are more easily used. The slotted line has the advantage that there is almost nothing that can go wrong with it that will cause it to give an incorrect reading of SWR, so its main use is as a laboratory instrument for checking the calibration of more user-friendly meters.

Information about the matching between line and load may be presented in several different ways. Another that is common in some circles is a specification of *return loss*. This is a measurement of the power reflected by the mismatch. If either the return loss or standing wave ratio for a particular connection is known, the other can be calculated, but the relationship is not a linear function (see Fig. 7.19). A short circuit or open circuit load that absorbs no power at all gives a return loss of 0 dB; a perfectly matched load is measured as infinite return loss. A common benchmark for small transmitters and transceivers is a standing wave ratio of 2:1. This is achieved when 89% of the power supplied by the line is accepted and radiated by the load, and 11% is reflected. There is a 9 dB difference between 11% of the power and 100% of the power, so an SWR of 2:1 indicates a return loss of 9 dB.

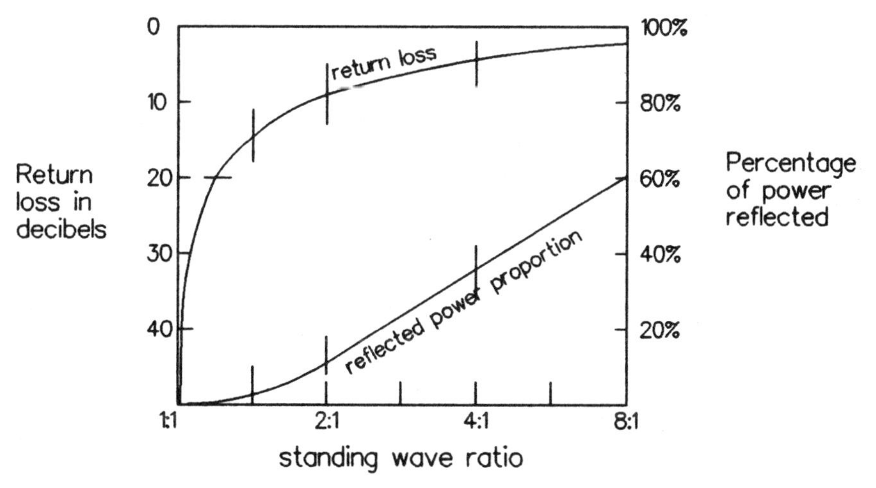

Fig. 7.19 *Graph showing return loss (dB), and percentage of reflected power, against standing-wave ratio*

Return loss is the factor of interest for television transmitter aerial systems and some broadband data services. Where the highest modulating frequency corresponds to a wavelength much longer than the length of line between transmitter and aerial (as applies to most transmissions of voice signals), there are opportunities to correct quite bad mismatches by adjustments at the transmitter end of the line. In the case of a television signal, however, that correcting signal cannot be used because it is time-delayed by the round trip down the line and back up again and would lead to the transmission of a ghost image. Television transmit aerial matching must be done by careful and finicky adjustments at the top of the tower with a return loss figure of about 40 dB being aimed for.

7.13 THE QUARTER-WAVE MATCHING SECTION

You will need to have a thorough understanding of Sections 7.11 and 7.12 before reading this section. It is a bit 'heavy' in the mathematical sense but it is worth wading through it because it describes a key factor in the design of many radiating systems.

When a transmission line is connected to a mismatched load, power is reflected and the phase of the reflected power at the point of the load is controlled by the resistance and reactance of the load compared to the characteristic impedance of the line. As the reflected power moves back along the line, its phase relationship with the incident signal changes in a cyclic fashion. There are points on the line where the relative phase is the same as the relationship that applies at the point of the load. These points are each a quarter of a wavelength apart.

The conditions that would be shown by measurements would not, however, be exactly the same as at the load. For instance, if the load is seen as a pure resistance of lower value than the characteristic impedance of the line, then the voltage measured at the load will be lower and the current higher than for the incident power, and the two signals will be in phase. At the point along the line one-quarter of a wavelength away from the load, incident and reflected signals will each have changed in phase by 90° but in opposite directions. The voltage and current will again show a resistive phase relationship, but at this point the voltage is higher and the current lower by the same amount as the mismatch at the load.

For instance, a pure resistance of 10 ohms connected as the load for a 50 ohm transmission line and then measured at a point $\frac{1}{4}$ of a wavelength back along the line will indicate the voltage, current, and phase of a 250 ohm pure resistance at the point of measurement (see Fig. 7.20).

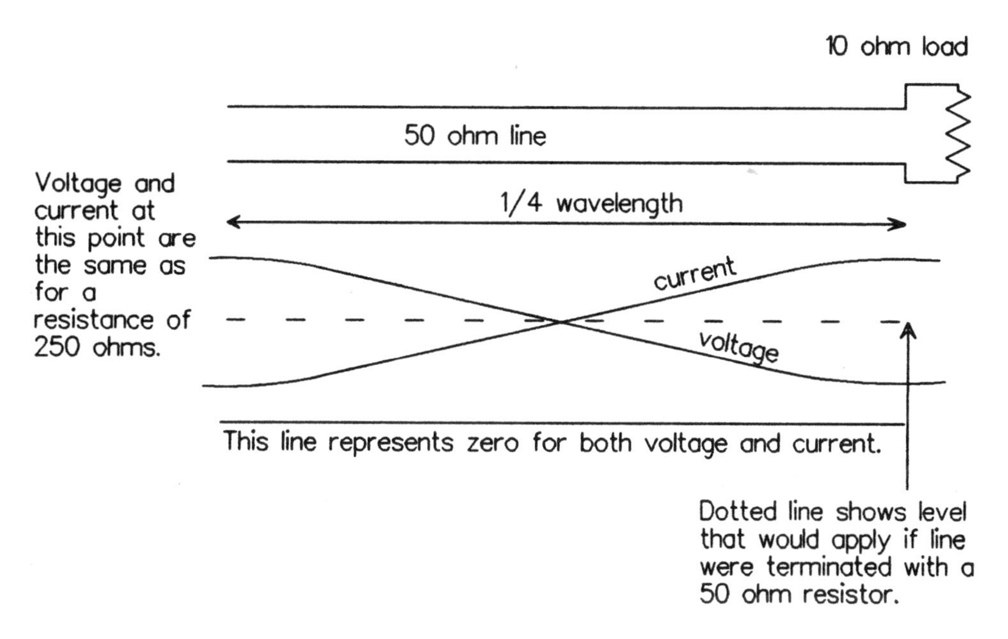

Fig. 7.20 *Diagram showing voltage and current conditions on a quarter-wave line with mismatched load*

If the load has reactance, the current and voltage will not be exactly in phase at the point of the load, and they will be out of phase to the same extent at the quarter-wave measurement point and the signal will behave as if the load were a reactance of the opposite type. A load that includes a small value of series inductance is seen at the quarter-wave measurement point as if it were a large value of capacitive reactance (small capacitance) in parallel with the resistive component.

Quarter-wave sections of line have an impedance transforming action which can give a very versatile low-loss form of impedance matching. They are very commonly used for that purpose in transmitting aerial arrays to adjust the system for 1:1 SWR on the line. One special case of interest is that where the quarter-wave line is terminated by a short circuit. At the measurement point it appears as an open circuit. A shorted quarter-wave section of line can be used as a tuned circuit in which the capacitor and inductance are in parallel. That feature is useful where a signal must be transferred to another point and reversed in phase in the transfer. Vertical collinear aerials may use this form of matching between sections.

Sections of line which are slightly longer or shorter than a true quarter-wavelength will show changes of phase and can be used to tune out reactance from a radiating element that cannot be made exactly the tuned length. The most common use of this principle is for stub matching. The arrangement shown in Figure 7.21 is one method of matching that can be used with VHF mobile transceivers. The initial installation is made with the stub too long, and then small amounts are cut off until a meter in the coaxial line shows minimum standing wave ratio.

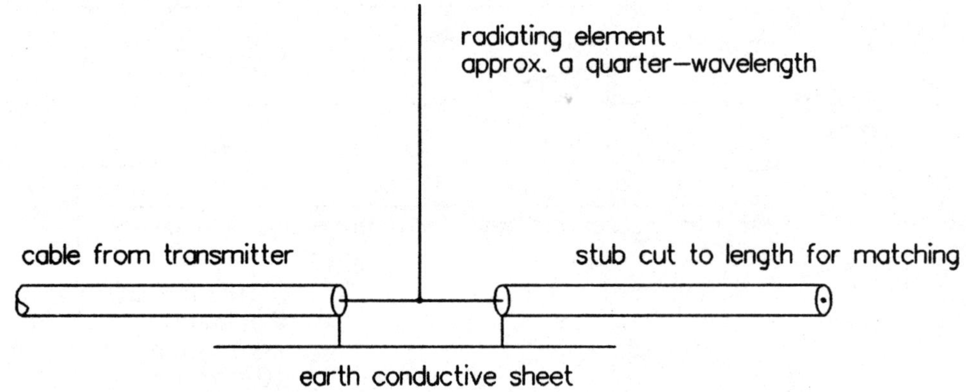

Fig. 7.21 *Whip aerial with stub matching*

CHAPTER 8

AN EXAMPLE OF EACH TYPE OF AERIAL

8.1 CHAPTER OUTLINE

The equipment described in this chapter is intended to show some practical applications of the principles laid down in Chapter 7.

Section 7.2, 'Initial classification', suggested grouping all radiating structures into these four classes:

1. resonant, omnidirectional
2. resonant, with directivity
3. non-resonant, omnidirectional
4. non-resonant, with directivity

This chapter lists the names of some of the common types of structures within each group and gives some detail of the construction of at least one of the listed types.

8.2 RESONANT, OMNIDIRECTIONAL

Note at this stage the slight difference in terminology between *omnidirectional* and *isotropic*. The true isotropic radiator is a theoretical beast that cannot exist in practice. If an aerial is designed for terrestrial use and gives even coverage for ground-based observers at all points of the compass with effective radiated power within about 1 decibel of the theoretical figure for an isotropic radiator, then its performance in all practical situations is that of an isotropic aerial. If a real aerial is described as 'isotropic', then it is only being considered in two dimensions and the term really means 'omnidirectional with unity gain'. Examples of resonant, omnidirectional aerials with close to unity gain are:

1. quarter-wave verticals;
2. electrically short verticals and mobile whips of many different types;
3. crossed dipoles.

The loop and ferrite rod with tuning-coil assemblies used in almost all portable broadcast receivers can, for most practical terms, be included in this description, although to be strictly correct they show polar diagram responses similar to that of a half-wave dipole. When you are receiving a local broadcasting station, the automatic gain control circuit of the receiver makes the response appear to be omnidirectional.

Aerials may be built with an omnidirectional response in the horizontal plane but with gain achieved by limiting and focusing the vertical response. The vertical collinear arrays, commonly used for UHF CB base stations and VHF/UHF broadcasting and television transmitting aerials, have that function.

8.3 THE QUARTER-WAVE VERTICAL

It is assumed that you have read and understood Section 7.4, 'The driven element'. Imagine that a half-wave dipole could be cut in half and one of the halves replaced with a very large sheet of conductive material. For radio waves the conductive material acts as a mirror, and to the outside world the structure looks exactly the same as a half-wave dipole. The body of the Earth can be made to look like a radio mirror so a quarter-wavelength of conductor placed vertically can be made to radiate very effectively (see Fig. 8.1). This arrangement matches well with signals designed for propagation by ground wave because the direction of maximum radiation is along the surface of the Earth in all directions.

Coupling of power to the quarter-wave vertical is somewhat changed from that of a dipole. The natural connection of a centre-fed dipole is a balanced two-wire line of about 70 to 78 ohms impedance. For the vertical, this is effectively split in half; one of the conductors is connected to earth and the impedance is halved. A coaxial cable with the outer conductor earthed and forming a shield for the live inner conductor is a natural companion to the quarter-wave vertical radiator.

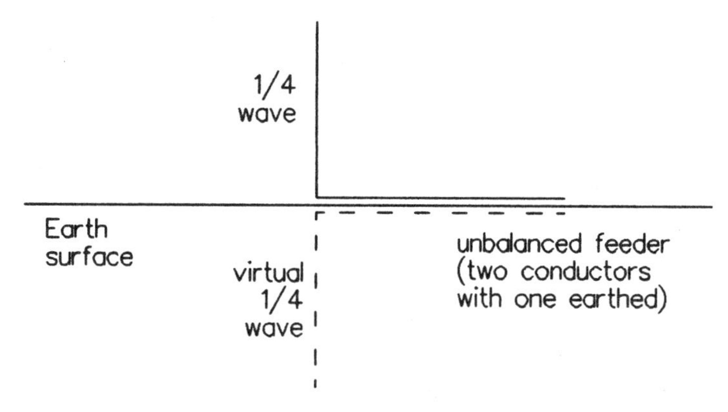

Fig. 8.1 *Quarter-wave element perpendicular to a reflecting surface*

8.4 ELECTRICALLY SHORT ANTENNAS

The principles outlined in this section apply mainly to two types of structure. One is the type of vertical aerial used on the VLF and LF bands; the other is a variety of designs of loaded whips used by mobile stations operating in the MF and HF bands. In a somewhat less restrictive form, the same principles are used by the 'stubby' antennas of VHF and UHF hand-held transceivers (including cellular phones).

This section deals with theoretical principles and leaves you to recognise the functions of the various physical elements of the structure.

The common factor of all radiators of this type is that the straight-line length available is limited by physical constraints to much less than a quarter of a wavelength. The starting point of the design of an electrically short radiator is that the free end must always be a point of high voltage and no current. A slight current can flow in the piece of conductor just a bit back from the end because electrons will pack in a bit tighter in response to the applied voltage. The woolly sheep analogy made in Section 7.4 is equally relevant here.

In a quarter-wave vertical aerial, the current flow is at a maximum at the base and reduces to nothing at the top. Radiation of power is dependent on current flow, so most of the radiation actually comes from the lower section of the conductor. When you attempt to shorten the structure by cutting off a section of the whip or tower, the electrical effect you have is to remove length from the base end which is where most of the current is flowing (Fig. 8.2).

The general design aim is to make a structure in which, for most of the available length, current can flow as if it were at the base of a quarter-wave section. For fixed installations, if altitude is limited but paddock space is

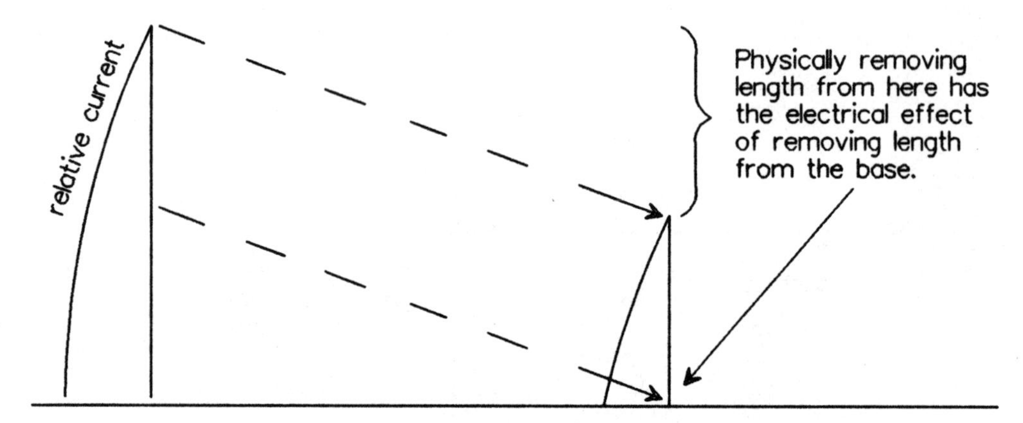

Fig. 8.2 *Diagram of current distribution for a quarter-wave vertical aerial and for a shortened aerial*

available, extra wire can be added horizontally; in some cases, enough length can be gained by this means to make the completed structure behave as a quarter-wave vertical. (See Fig. 8.3.)

When paddock space is also restricted, conductors can be added in several different directions to make what is called a *capacity hat*. Some of them look for all the world like a gigantic version of a rotary clothesline. The aim of all these structures is to give the current somewhere to go when it gets to the end of the vertical radiating section. The diagram of the MF broadcasting transmitter aerial in Figure 7.1 shows a quarter-wave vertical shortened by the addition of a capacity hat.

To the electrical circuit, a quarter-wave vertical behaves as a series tuned circuit connected in series with a resistance equal in value to the radiation resistance (Fig. 8.4). The design aim is that most of the power supplied by the transmitter should be dissipated by the radiation resistance as radiated signal.

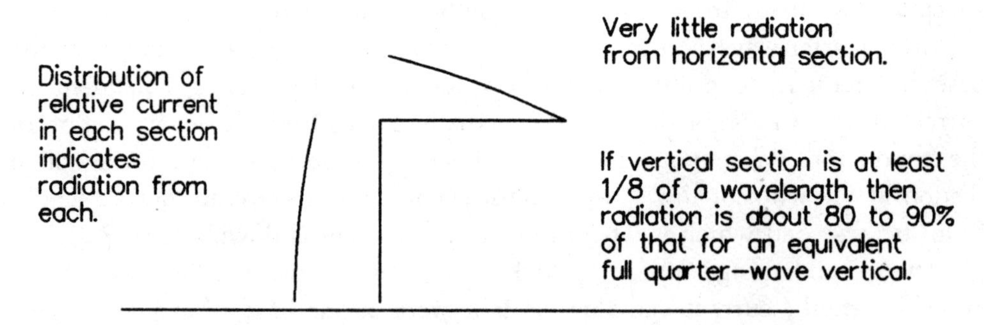

Fig. 8.3 *Extra wire added horizontally when altitude is limited but paddock space is available*

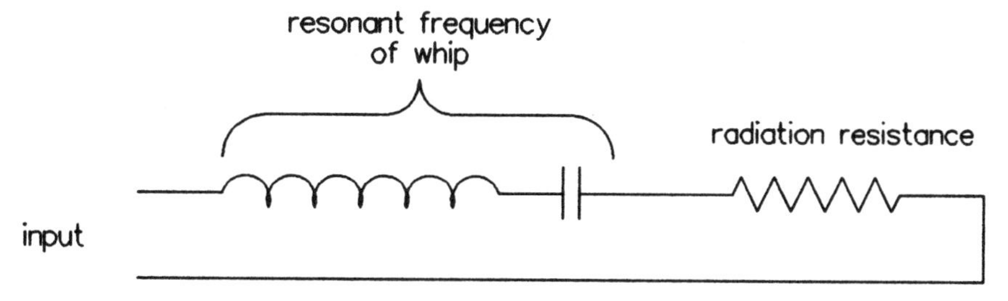

Fig. 8.4 *Diagram of the equivalent circuit of a quarter-wave vertical*

When the structure is shortened, it is detuned and the circuit behaves as a resistance in series with a capacitive reactance which impedes the applied voltage and limits current flow (see Fig. 8.5).

The capacitive reactance can be tuned out by adding the correct series inductance. This then allows power to be coupled to the radiation resistance without the limiting effect of the capacitive reactance. This series inductance element is called a *loading coil* (Fig. 8.6).

A side effect of reducing the length of conductor in the radiating section is that coupling between the current and the radiation field is reduced. The practical effect of this reduced coupling is a drop in the ohmic value of the radiation resistance. Small changes do not matter because you can still couple power to it by adjusting the matching ratio (same effect as changing gears in a motor car). For large reductions in radiation resistance, however, there are two factors which make a short whip with a loading coil difficult to use:

1. The workable bandwidth of the system is reduced which can mean that only a few channels out of a band are usable.
2. The power applied is shared between the radiation resistance and whatever there is in resistance to earth; so for very low values of radiation resistance, most of the power is converted to heat.

To some extent radiation resistance can be raised by placing the loading coil some distance up the whip. For a fixed installation with a capacity hat, the coil

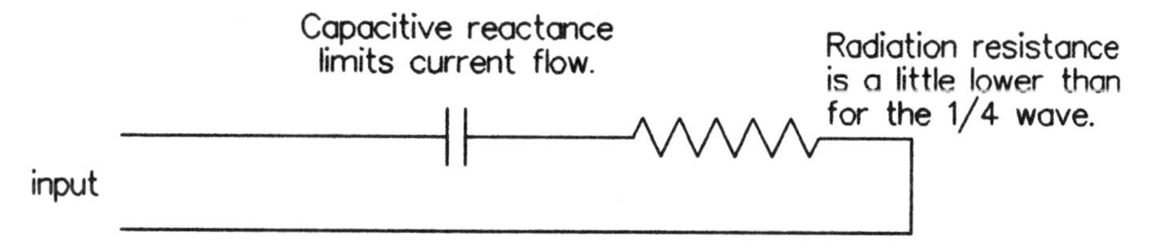

Fig. 8.5 *Diagram of the equivalent circuit of a shortened vertical*

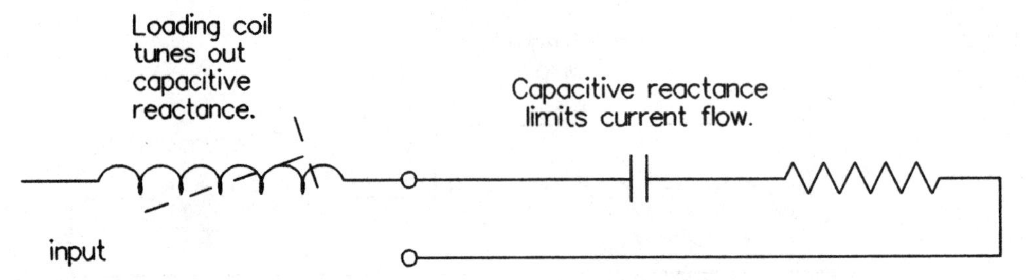

Fig. 8.6 *Capacitive reactance is tuned out by adding an inductance element called a loading coil*

can be effective even when placed immediately below the hat. However, this is not an option for a mobile whip.

There is a practical limit to how high up the whip the coil can be placed because it too requires that some current flow through it which means that there must be some length of wire above it for the current to flow into. A helically wound whip is the logical design extension of the whip and loading coil principle. For the wound whip, the loading coil is distributed along the length of the whip so maximum use is made of whatever length is available. (See Fig. 8.7.)

The general field of electrically short whips for mobile and portable use is one where the boundaries of the state-of-the-art technology are being actively

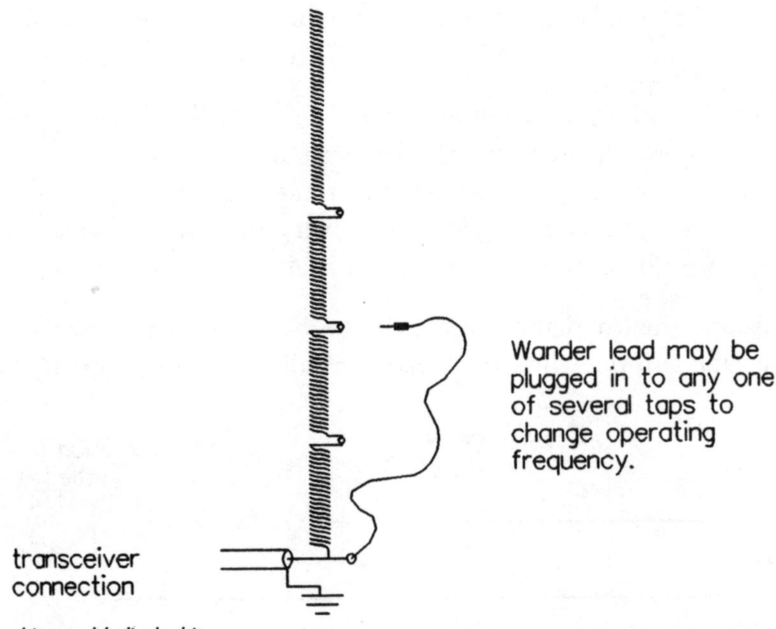

Fig. 8.7 *A multitapped helical whip*

extended, and there are new designs and detail improvements to old designs being made all the time.

8.5 ANTIFADING RADIATORS

This section is not essential to the main thread of the story so it can be omitted if it seems too involved; however, it does give a good practical example of how the principles of electronics and radiophysics can be trimmed and manipulated to achieve performance in particular directions.

If you live near to or have ever visited any of the state capital cities, you have probably seen one or more of a particular type of large object, commonly between 150 and 200 metres tall, which can be seen prominently on the skyline from several kilometres away. Most people, however, have probably never seen one close up—high-power transmitter stations do not fit well with the genteel values of concentrated suburban living and so they tend to be stuck on out-of-the-way corners of ground that no one wants for anything else. Section 6.4 mentions the *antifading radiator* and gives an outline of the purpose of its design. Its principal use is for MF broadcasting in association with high-power transmitters intended to give a regional or state-wide coverage.

An antifading radiator could be mistaken at a passing glance for a quarter-wave vertical with a capacity hat. If, however, you estimate the height of the structure and calculate the working frequency, you will find an answer that is between a half and a third of what is actually used. It is the components at the sectioning point that make the structure into an antifading directional aerial. (See Fig. 8.8.)

The outer ring of the capacity hat is an end point which must be a point of high voltage and no current. The length of conductor from there to the sectioning point is an electrical quarter-wavelength. At the base where power is fed to the aerial, practicality demands that the electrical input looks to the feed circuit very similar to a quarter-wave vertical; that means it must be a point of high current and low voltage.

From the base to the sectioning point, the length of conductor is somewhere between a quarter and a half of a wavelength. There are lumped reactive components (coils and capacitors) at the sectioning point which make the whole thing workable. The required directivity pattern is achieved by trimming and juggling the exact lengths of conductors and the values of the components at the sectioning point.

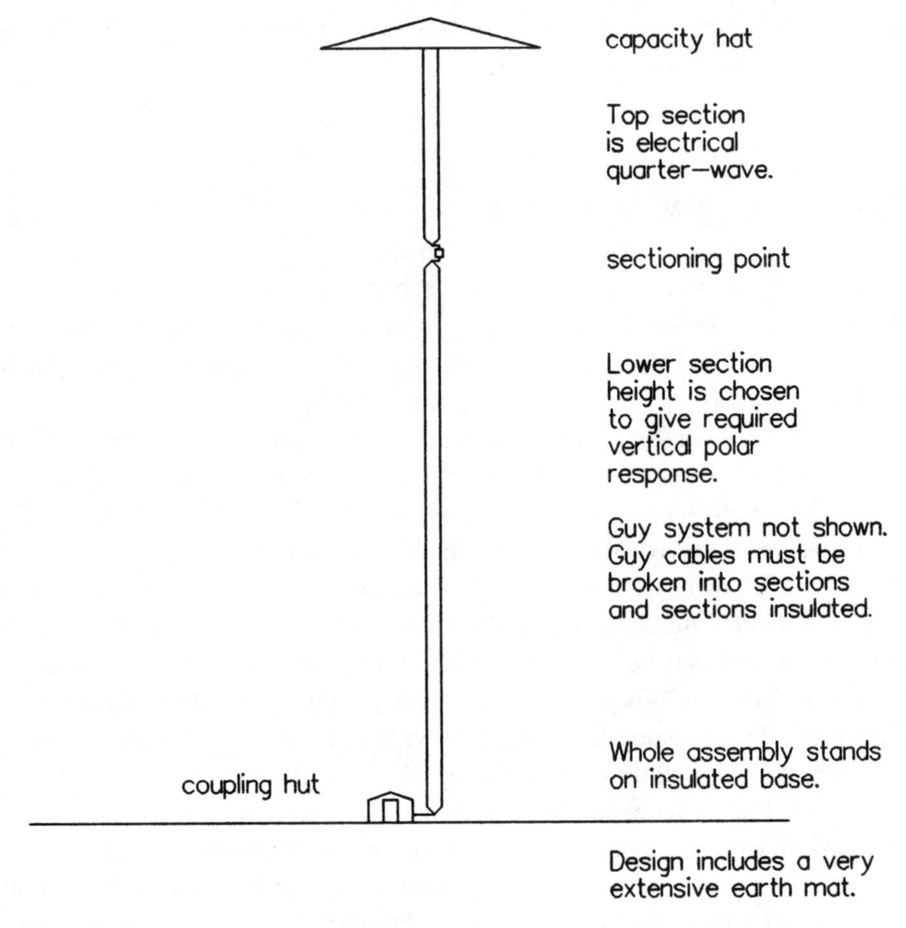

Fig. 8.8 *The design principles of an antifading radiator*

If the lower section of conductor were an exact quarter-wavelength, its top (at the sectioning point) would be a high impedance point and would be difficult or impossible to match to the lower impedance input of the upper section. However, if matching were possible and phase relationships of the two sections could be kept within acceptable limits, then radiation from the two sections would combine to increase the signal at low vertical angles and reduce it at high angles, reducing the sky wave at the distant receiver which is exactly what is being designed for.

If the lower section of conductor were a full half-wavelength, impedance matching would be easy; impedance conditions at its top would be exactly the same as at the feed point which would match nicely to the low impedance input of the upper section. Lumped components at the sectioning point only need to adjust relative phase between the two sections to achieve a significant increase in ground-wave signal strength.

The catch is that the vertical polar diagram of a radiator with a total electrical length of three-quarters of a wavelength shows a null at about 60° to 70° of vertical angle and a quite pronounced second lobe of radiation at higher angles. There are some circumstances where that secondary lobe could appear as an unwanted sky-wave signal which would actually reduce the distance to the first point at which sky wave and ground wave are of equal intensity. The source of this null and minor lobe pattern is described in Section 7.10, 'Major and minor lobes'. When electrical length is reduced below three-quarters wavelength, the angle of the null moves towards vertical and the total energy radiated into the second lobe is reduced.

Thus the final design of an antifading radiator is a compromise between several ideals. Total electrical length is adjusted so that energy in the secondary lobe is nil or very small. The null, if it exists, is aimed at a vertical angle chosen to reduce the sky wave where its presence has most effect, and lengths of conductor in each section are trimmed so that the lumped components in the sectioning point have some impedance matching and some phase adjustment effects but not an impossible task in either direction.

8.6 RESONANT, WITH DIRECTIVITY

The common types of aerials in this class are the *Yagi*, *corner reflector* and *screen-backed collinear array*. Any of these may be horizontally or vertically polarised.

There are many designs described by many different names which can be shown to use the basic principle of one or more of those three types with added facilities to give the finished structure extra capabilities. Features such as broadband or multifrequency operation or electrical slewing of the beam can be designed in if the need arises. For instance, almost all television receiving aerials are basically one of the above three types with modification for operation over the required broad range of frequencies.

8.7 THE YAGI

When a half-wave dipole driven element (as described in Section 7.4), a reflector and a director are mounted in the same plane (Fig. 8.9), the effects of the two

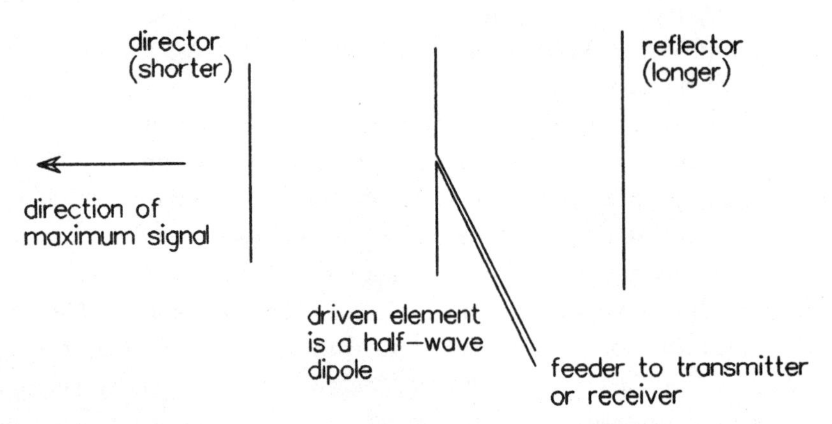

Fig. 8.9 *Yagi—an aerial with all elements mounted parallel and in the same plane*

parasitic elements combine to provide an increased gain in the strongest part of the beam.

For an arrangement such as this, the two figures of performance that are of major interest are the forward gain in the centre of the major lobe of the beam and the front-to-back ratio which specifies the signal in the exactly opposite direction. In other than these two directions, polar diagrams for a small Yagi show similarity to those for a half-wave dipole.

The Yagi has a very flexible design. Actual figures for forward gain and front-to-back ratio can be varied quite widely by small changes in lengths of the elements and their spacing. Spacing also affects the impedance seen by a signal arriving on the transmission line, so within limits can be used for matching of power transfer. Defining an effective and commercially repeatable design may involve the balancing of several partly conflicting requirements and can be an involved process even for something as simple as a three-element Yagi.

A Yagi can be made with more than three elements. Usually only one reflector is used but several directors will give a useful increase in concentration of the beam in the forward direction. The practical maximum is usually found to be about 8 to 10 directors which may give up to about 15 dB forward gain. Limiting factors are leakage of signal in other directions due to minor lobes, and the limited effectiveness of the single reflector at the rear. Figure 8.10 shows the Yagi polar diagrams.

8.8 COMBINING SEVERAL DRIVEN ELEMENTS

With the correct arrangement of feed lines, it is possible to feed power to several driven elements in combination. If you do that you affect the directional properties of the completed structure.

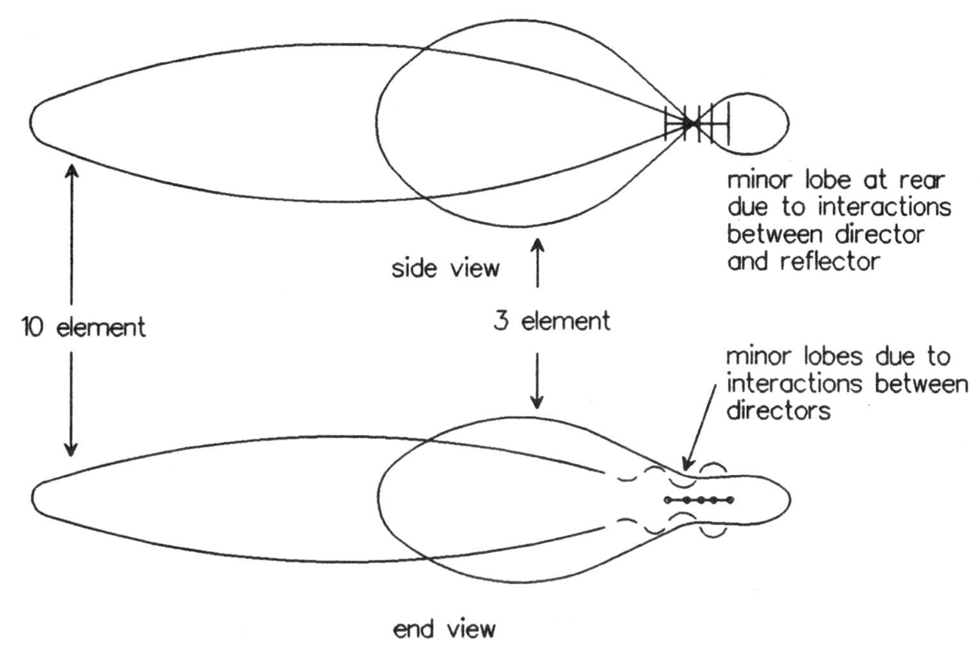

Fig. 8.10 *Polar diagrams for a three-element and a ten-element Yagi*

If two half-wave dipoles are mounted parallel and fed with power in phase and spaced half a wavelength apart, there will be directions in the plane at a right angle to the dipoles along the line joining the centres in which the signals cancel so there is no radiation. At a right angle to that line, the signals from the two dipoles add and radiation is increased. (See Fig. 8.11.)

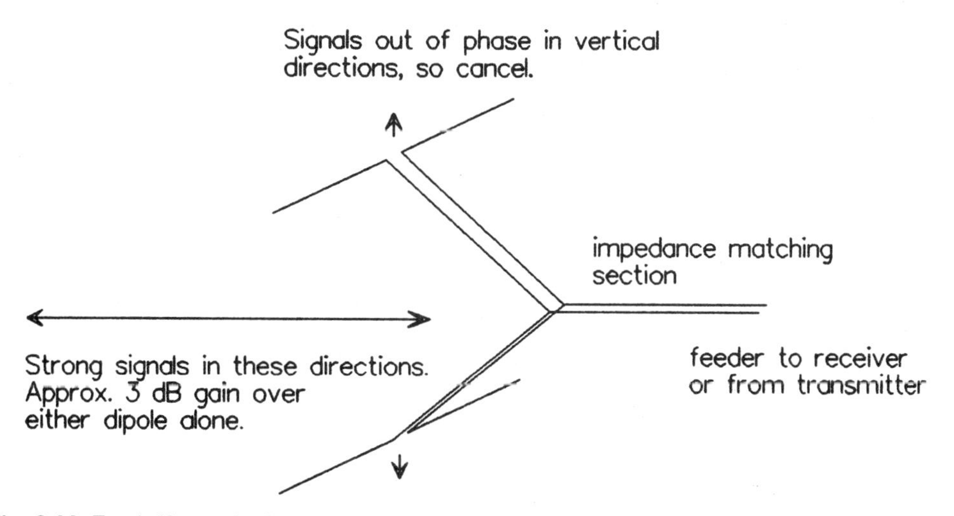

Fig. 8.11 *Two half-wave dipoles mounted parallel and in phase*

Dipoles can be placed either side-by-side or end-on, and quite large high-gain assemblies can be created by placing dipoles on a grid pattern in front of a reflecting screen. These are generally named *screen backed collinear arrays*. The maximum gain available is limited only by the size of the structure and the number of dipoles used; as a rule of thumb, the gain is raised about 3 dB each time the number of elements is doubled.

8.9 SCREENS AND CORNER REFLECTORS

If a half-wave dipole driven element is placed a quarter of a wavelength away from a large flat sheet of conductive material, the sheet of conductor reflects all the signal in its direction and has the effect of adding the reflected signal to that which started out going away from the sheet. The distinguishing feature of that arrangement is that almost no radiation gets through or past the sheet, and so aerial systems based on backing screens have a very high front-to-back ratio. (See Fig. 8.12.)

It has been found in practice that a solid sheet is not necessary, and practical screen-backed structures generally use a row of rods, wires or mesh running parallel to the driven element in the place where the solid screen would have been. If the rods are spaced closer than about $\frac{1}{10}$ of a wavelength to each other and each one is at least 30% to 40% longer than the driven element, they will have exactly the same effect on the signal as a solid sheet of similar outline dimensions. A single dipole placed before such a screen can have a forward gain of up to 6 dB, and front-to-back ratio may be over 20 dB.

With a single driven element, the forward gain may be increased by bending the screen into a corner (Fig. 8.13). The sharper the angle of the corner, the higher the forward gain. In theory, the gain can be made any required amount just by making the angle of the corner sufficiently small. In practice, structures

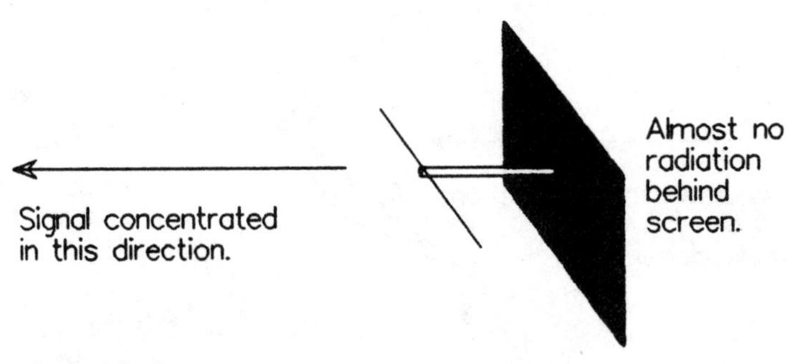

Fig. 8.12 *A screen-backed dipole*

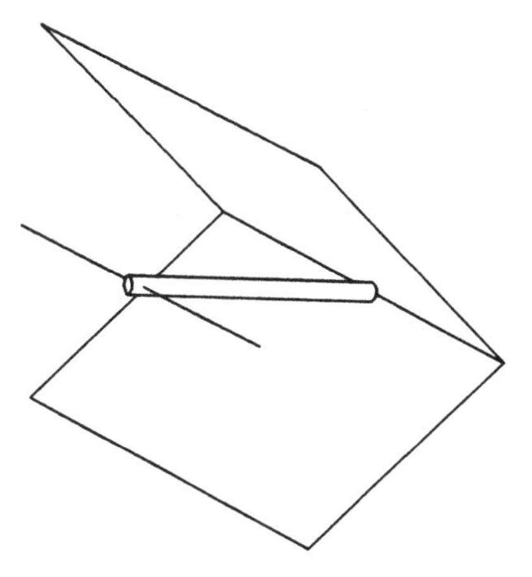

Fig. 8.13 *A corner reflector aerial*

with corners much sharper than 90° become unwieldy with very large areas of screen required. A corner angle of 90° with screens of the correct size gives forward gain of about 9 dB.

8.10 NON-RESONANT, OMNIDIRECTIONAL

Aerials in this class are not often used for transmitting but there are a couple of types that are commonly used for receiving. The whip aerial fitted to most passenger cars for reception of MF broadcasting fits into this classification. The Discone, which has become more common since the advent of scanning receivers, is also a non-resonant omnidirectional device (Fig. 8.14). Discones can be used for transmitting but the need for combining broadband response with omnidirectional coverage does not arise very often.

8.11 THE CAR RADIO RECEIVER

For signals in the MF band (and the low end of the HF range), the maximum possible clarity of reception is limited by background noise from outside the equipment. Sources such as thunderstorms and industrial uses of electricity will always place a limit on the best available signal-to-noise ratio. The transmitter is required to generate a sufficient field strength over all its primary service area to overcome the effect of local noise sources, and so high-power transmitters and high-efficiency radiators are used. At the receiver, however, high efficiency of signal collection is of little value. The signal only needs to be a few microvolts

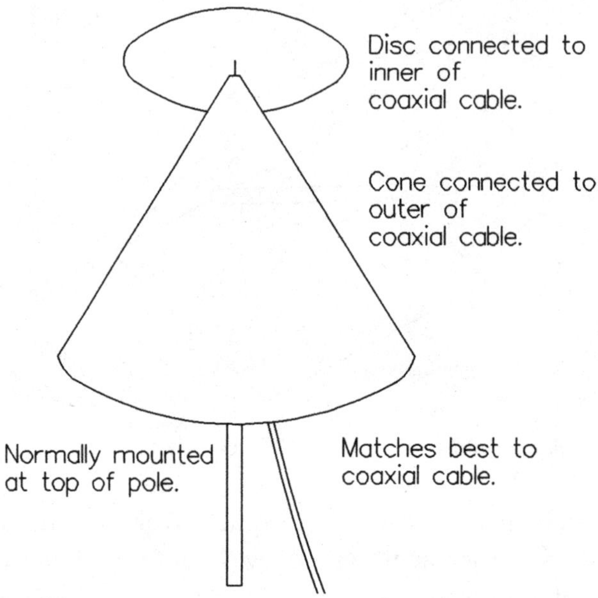

Fig. 8.14 *The Discone aerial*

at the input terminals to be strong enough to mask all the internal noise of the receiver.

Section 8.4, 'Electrically short antennas', described how a whip that is shorter than a quarter-wavelength looks to the driving circuit as if it were a resistance equal to the radiation resistance in series with a capacitance. In the receiving mode, a voltage appears across the radiation resistance (Fig. 8.15) which is a function of the length of the whip in metres multiplied by the field strength of

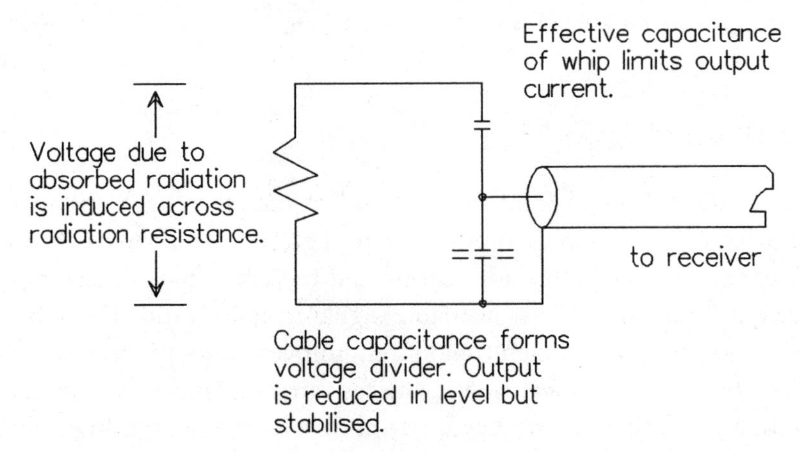

Fig. 8.15 *Schematic diagram of the equivalent circuit of a car radio receiver aerial*

the signal in volts per metre. This voltage is fed through the small series capacitance (only a few picofarads for a 1 to 2 metre whip) and a short length of shielded cable to the receiver electronic circuits. The cable has its own capacitance which provides a voltage divider effect on the signal. Received signal is small and the impedance of the aerial as a source is very high. So the power available is minute, but as long as it is sufficient to mask the internal noise of the receiver, that is all it must do.

The big advantage of this arrangement is that the output signal is almost constant over the 3:1 frequency range of the MF broadcasting band. There are no stray resonances, and external factors such as being parked next to a galvanised iron shed have very little effect on reception.

8.12 NON-RESONANT, WITH DIRECTIVITY

There are two radically different types of structure that fit this description. Large (in terms of wavelength) sheets of conductive material may be shaped into either parabolic reflectors or exponential horns to direct a highly focused beam in a particular direction. There are a number of aerial types that use the principle of the *terminated long wire*. Two of the most common constructions that use this principle are the *rhombic* and the *inverted vee*. The design of these for good directivity is a lot more complex than the simple appearance of the finished structure would suggest.

A well-designed rhombic can give high gain (up to 25 dB has been reported) over a very broad frequency range. Typically, a 3:1 or 4:1 ratio between highest and lowest frequencies is possible. To get performance of this order, however, each leg must be many wavelengths long. They require a lot of paddock space and generally cannot be made steerable.

8.13 DISHES AND HORNS

For very high gain (above 25 dB), and particularly if associated with wavelengths shorter than 300 mm, the number of elements required for either a parasitic or a screen-backed array would make the physical structure unwieldy. It is much simpler to make the radiating structure from a flat sheet of conductor formed into a particular shape. There are two ways this can conveniently be done.

The flat sheet can be formed into a parabola-shaped reflector (Fig. 8.16) like a spotlight reflector. This shape is technically called a *paraboloid of revolution*.

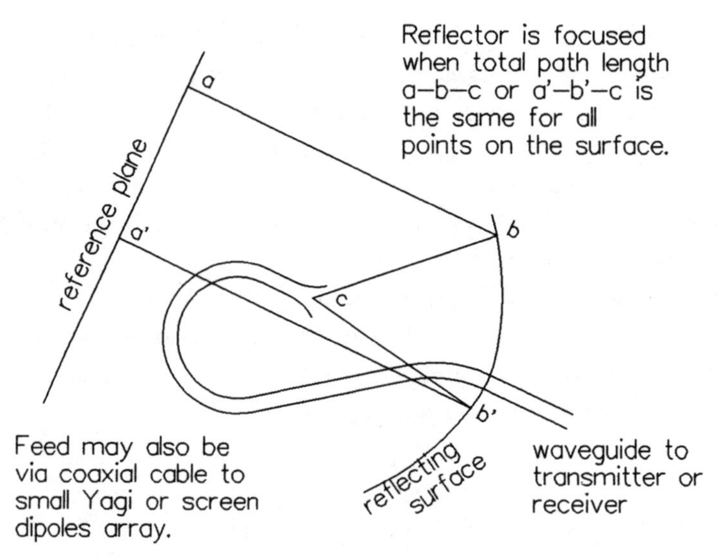

Fig. 8.16 *Parabolic reflector*

There is a tiny, fairly simple antenna placed at the focus point where the globe would be in a spotlight. Antenna gain depends on capture area which is closely related to the physical measurement of area across the face of the dish. Gain of 40 dB is not difficult to achieve, and in some laboratory tests, gain of up to 70 dB has been reported.

When the feed line is a waveguide rather than a cable system, a very simple method of achieving antenna gain is to simply flare out the end of the waveguide into a horn (Fig. 8.17). The aim is that the wavefront across the face of the horn

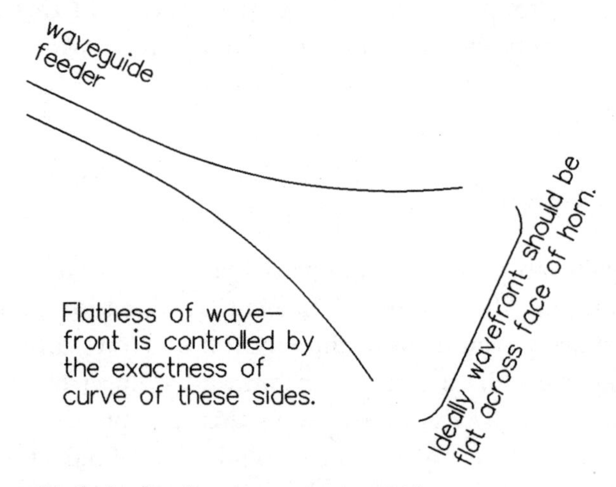

Fig. 8.17 *End of waveguide feeder flared out into an exponential horn*

should be flat, and if that is achieved, then antenna gain is directly related to capture area which can be defined with a tape measure. The principle of the horn radiator is very simple, but practical realisation of it may be a bit tricky. Ensuring that the wavefront really is flat is the critical part, and matching to the impedance of the waveguide without standing wave reflections is also important. To aid that, the sides of the horn should follow an exponential curve. When all is working properly, the antenna gain of a horn should be about the same as that of a parabolic reflector of equivalent face area.

CHAPTER 9

HEARING THE MESSAGE

This chapter is written assuming you have a good working knowledge of all of Chapters 2 and 4.

9.1 THE COMMON REQUIREMENTS OF ALL RECEIVERS

The receiving aerial or antenna will pick up a proportion of any electromagnetic energy that is going past and present it to the input of the receiver as a voltage that varies with time. There is little preselection in this process—the input voltage will contain a multitude of intelligent signals and a lot of other energy components that are not from intelligent sources. A primary function of a receiver is to select out the one required signal and discriminate against all others.

The other functions of a receiver are amplification and detection of the modulation and then it must present the signal in a form suitable as an input to a device that can use the information. The receiver is normally required to amplify the required signal. Input voltages in the few microvolts range are fairly normal—too weak to be directly used for any practical purpose but they can control the input of an amplifier.

After the signal is selected and amplified, the intelligent modulation must be revealed. This process is called either *detection* or *demodulation*. The designer of the receiver must make some presumption about what type of modulation is expected and build in the appropriate circuit. Finally, the signal must be presented in a usable form to another device which could typically be a loudspeaker, a television picture tube, a computer or the hybrid at the input of one line of a telephone exchange.

This chapter contains information which is mainly about electronic circuits. A certain amount of basic electronics and electrical theory is unavoidable. Be patient while you wade through it. If it seems too heavy to chew all in one lump, you may jump forward to the summary after Section 9.9, then branch back to the relevant section for more detail on each block of the diagram.

9.2 TUNING AND FILTERS

There are three types of passive electrical components which will have an impeding or limiting effect on current flow when a voltage is applied. They are *resistances*, *inductances* and *capacitances*. (In some cases, discrete components that are designed to provide capacitance may be named *condensers*.)

In a resistance, current flow is proportional to the applied voltage and inversely proportional to the value of the resistance. This applies equally whether the applied voltage is steady as from a battery, or is alternating, or is any combination of the two. Figure 9.1 shows a graph of driving voltage and resulting current for a 0.2 ohm pure resistance.

When the voltage is alternating, the current through a pure resistance is in phase with the voltage and is the same for all frequencies of alternation. (There are practical side effects that put limits on the workable frequency range of real resistors, but within the working frequency range, that principle holds.)

Because current and voltage are in phase, resistors consume power and convert the energy into heat. The power is calculated by multiplying the electrical force in volts by the current flow in amps to give a figure for power in watts.

A different type of impedance is *inductive reactance*. Chapter 3 describes how energy is stored in a magnetic field and explains that the effect of that energy storage is to slow the build-up of current when voltage is first applied and to tend to prolong the current flow when the voltage is removed. For an alternating

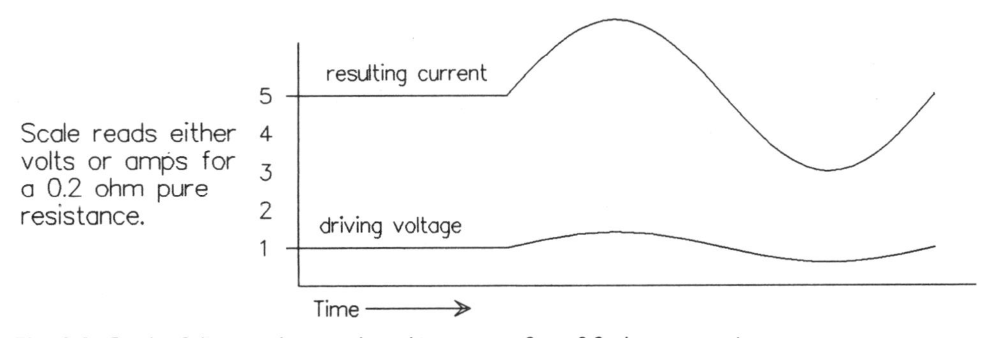

Fig. 9.1 *Graph of driving voltage and resulting current for a 0.2 ohm pure resistance*

voltage, this energy storage has the effect of applying an impedance to current flow which has some of the effects of a resistance but also some very important differences. As with resistance, the current flow is proportional to applied voltage and inversely proportional to inductance *but* it is also dependent on frequency. Inductive reactance is zero for direct current (zero frequency) and rises in proportion to frequency of alternation. (See Fig. 9.2.)

The other important difference is the phase relationship. If the alternating voltage is a sine wave and the circuit under examination has only inductance and no other forms of impedance, the current waveform will be the same shape but displaced 90° later in the cycle. This is generally described as lagging in phase by 90°. Because of this phase displacement, power is drawn from the electrical circuit for one half-cycle and returned to it on the next; the average power flow over any time longer than a few cycles is zero.

The third type of impedance is capacitive reactance. Electrical capacity or *capacitance* is a measure of the storage of energy in electric fields between two plates charged to different voltages (also described in Chapter 3). In the case of capacitance, the current flow is dependent on voltage, on the size of the capacitance and also on frequency, but in this case, current flow for zero frequency is zero and it rises in proportion with frequency. This means that the impedance measured across the component falls with increasing frequency.

There is also a phase difference between current and voltage in the capacitive circuit. If the alternating voltage is a sine wave, the current flow will also be a sine wave displaced in phase 90° before the timing of the voltage wave (Fig. 9.3).

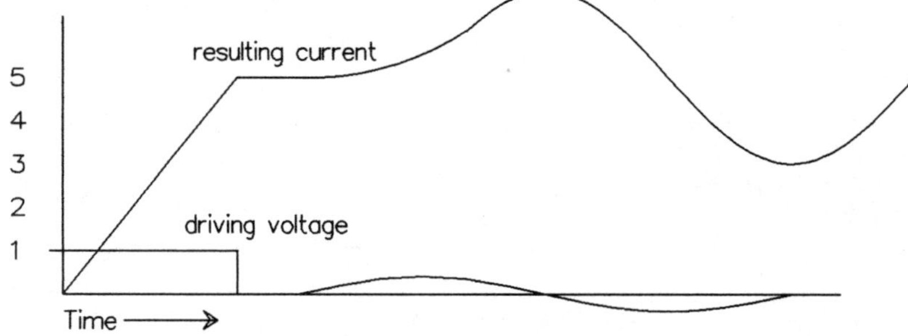

Fig. 9.2 *Graph of driving voltage and resulting current in an inductive reactance*

In a capacitive reactance a sine wave driving voltage will also produce a sine wave current but in this case the waveform is displaced by 1/4 of a cycle leading the voltage.

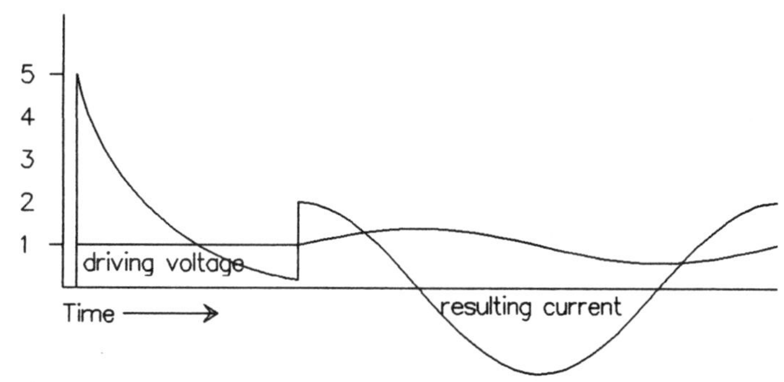

Fig. 9.3 *Graph of driving voltage and resulting current in a capacitive reactance*

A significant new phenomenon arises when inductive and capacitive components are combined in a circuit which carries an alternating current. In a series circuit, the same current flows through all components, and voltages are developed across each in response to the current flowing through each impedance (see Fig. 9.4).

Each voltage is at 90° phase angle to the current but in opposite directions. There is a total of 180° of phase difference between the voltages across the two reactive components which means that, in effect, one voltage is subtracted from the other. The result for the external circuit is that the combined impedance is the difference between the two reactances, and at one particular frequency when the reactances are equal, the only impedance left in the circuit is the small

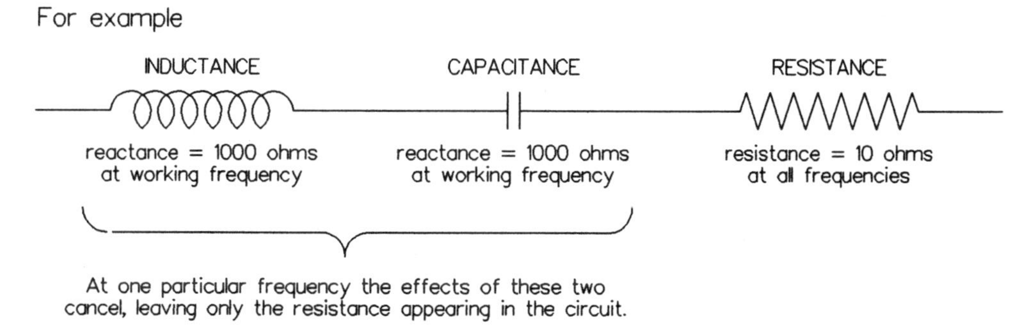

Fig. 9.4 *Inductance, capacitance and resistance in series*

residual resistance due to the wire of the coil and possibly some resistive losses in the capacitor.

In the example in Figure 9.4 if the frequency is changed by 1% (say 1% higher), the inductive reactance becomes 1010 ohms and the capacitive reactance 990 ohms (approx.). This leaves a difference of 20 ohms which combines with the 10 ohms of the resistance by the rules of Pythagoras's theorem to give 22.36 ohms total impedance.

In a parallel circuit, the same effect applies but in this case the same voltage is applied to both and the currents subtract from each other. At the particular frequency when reactances are equal, the residual current is theoretically zero although there is a small residual current due to resistive losses in the reactances.

A *tuned circuit* is formed whenever inductive and capacitive reactances are combined. In Figure 9.5, graphs 1 and 2 show inductive reactance and capacitive reactance respectively against percentage of resonant frequency, and graph 3 shows inductive and capacitive reactances combined in a series circuit against percentage of resonant frequency. The resonant frequency is the frequency at which the two reactances are equal. In a tuned circuit, energy is stored in the form of an oscillation that sloshes back and forth between the inductance and the capacitance. You may think of the inductance as the electrical equivalent of inertia and the capacitance as the electrical equivalent of elasticity. In a mechanical device such as a tuning fork, the combination of inertia and elasticity stores energy as an oscillation at a particular frequency, and in the electrical circuit, the combination of inductance and capacitance does the same.

A tuned circuit combined with an amplifier (such as a transistor or integrated circuit) (see Fig. 9.6) will give the facility to amplify signals in a small range of frequencies only, as indicated in the graph in Figure 9.7.

A single tuned circuit by itself does not give sufficient effect to select out a single channel; there are always stray impedances and coupling mechanisms which limit the ability to reject unwanted signals to a maximum of about 30 dB for each circuit. However, the combined effects of several amplifying stages and several tuned circuits can give whatever degree of selectivity is needed.

The performance required of course varies with the use of the system. A mantle receiver for local stations on the MF broadcasting band may have two amplifying stages and four or five tuned circuits and give all the performance ever required of it. A VHF or UHF radio-communications receiver would require five or six amplifying stages and ten to twelve tuned circuits to give acceptable service for its purpose.

The degree of selectivity of each tuned circuit can be controlled by varying the loading (ratio of reactance to resistance) on it. High selectivity has two

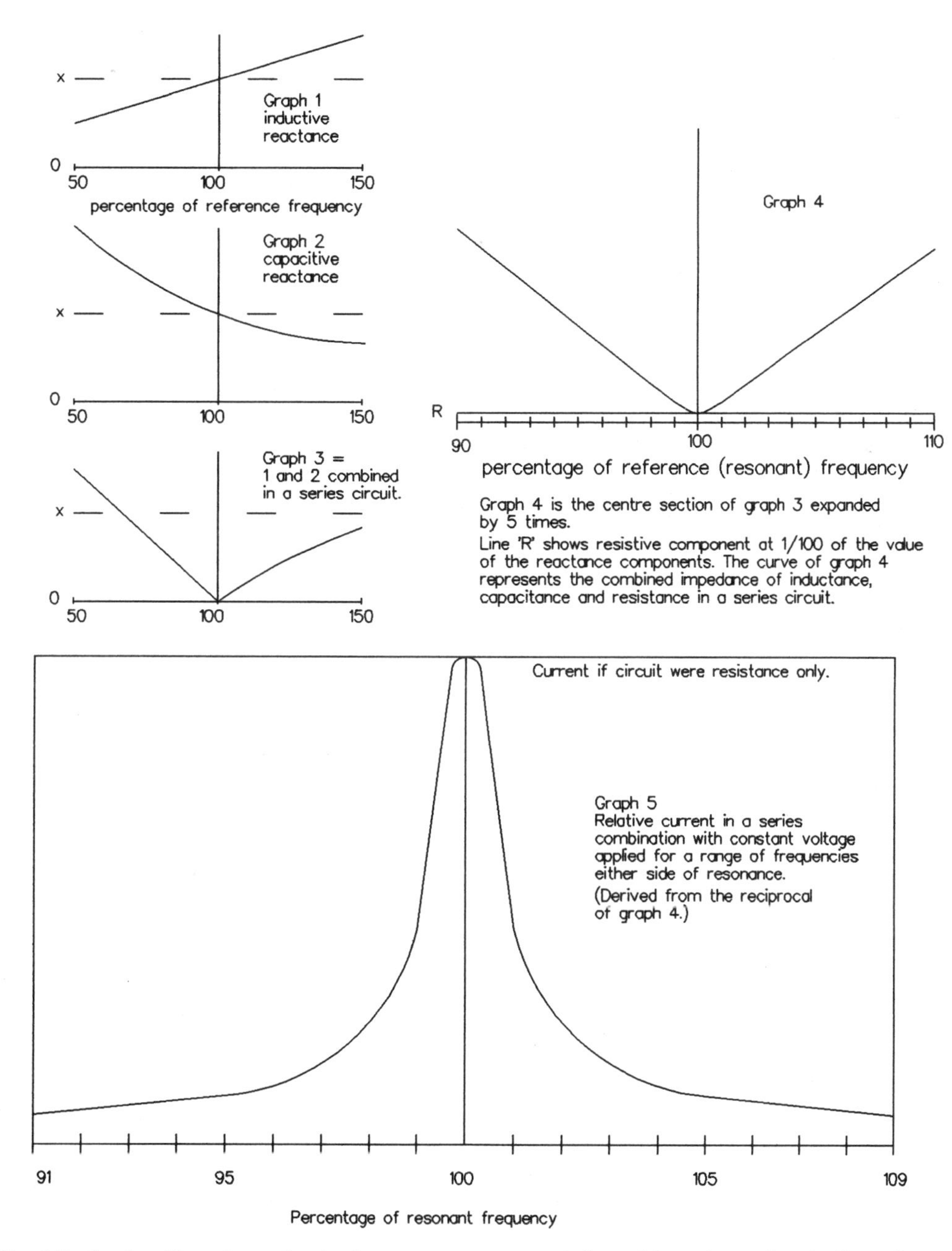

Fig. 9.5 *Graphs of impedance showing how a resonant system is formed from the combined effects of inductance and capacitance*

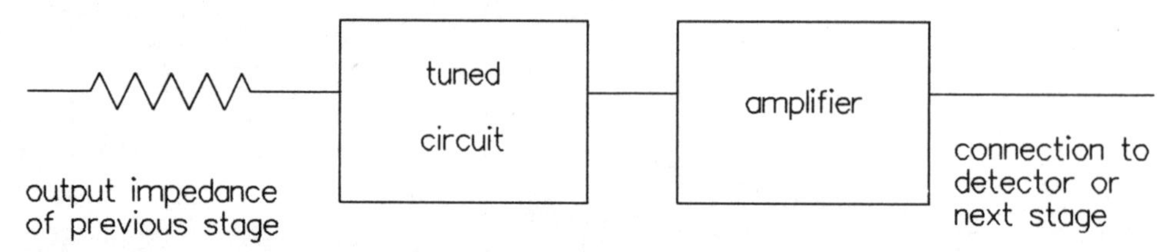

Fig. 9.6 *Block schematic diagram of a tuned amplifier*

effects: it gives good ability to reject signals on closely adjacent channels, and it also limits the ability of the receiver to detect the sidebands associated with the highest frequencies of the modulation. When tuned circuits are operated individually, the frequency-response graph will always be a graph of the same shape—just wider or narrower as the loading is varied (see Fig. 9.8).

For really high performance, what is needed is a frequency response that is flat over the low attenuation section close to the centre frequency, but rapidly moves to very high attenuation as the frequency is tested in the range where adjacent channel signals may be expected. This more tailored shape (see Fig. 9.8) is achieved with a more complex combination of reactive components called a *filter*. Filters are usually manufactured to have a particular centre frequency, a set bandwidth to cope with the expected range of sidebands, and a *shape factor* which indicates the degree to which the high attenuation sections of the frequency response curve are wider than the intended passband. Crowding of channels on the communication allocations of the bands generally requires that, for high

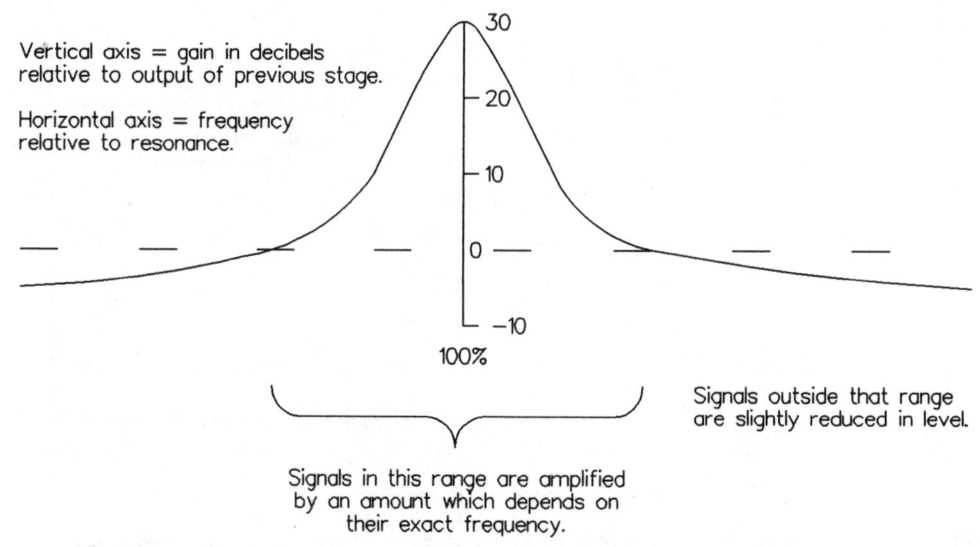

Fig. 9.7 *Graph of the output voltage versus frequency response of a tuned amplifier*

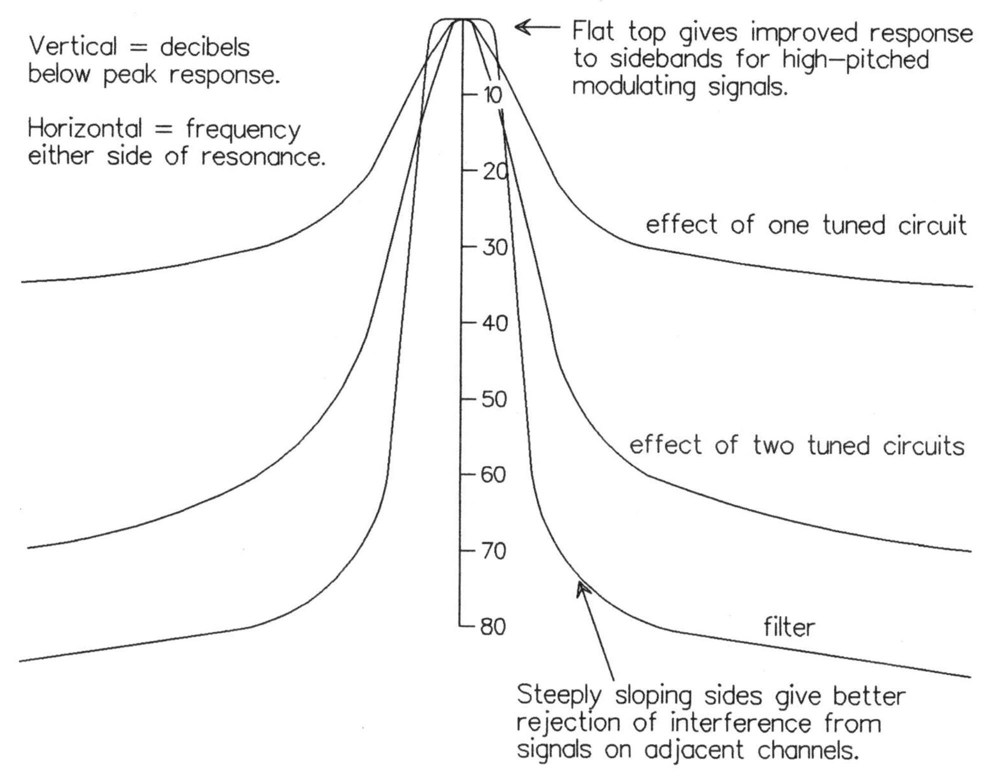

Fig. 9.8 *Frequency-response graphs comparing effects of tuned circuits with that of a filter*

performance, a receiver's selectivity is mainly provided by a filter rather than just a sequence of tuned circuits.

9.3 FREQUENCY TRANSLATION

As a step in practicality, the extra complication of a *frequency changer* stage is usually added to make the rest of the receiver simpler. Filters of the type described in Section 9.2 are always permanently tuned to a particular channel, so if the receiver must operate on any other frequency, that incoming frequency and the range of the spectrum either side of it must be translated so that the wanted signal is within the passband of the filter.

The process of frequency translation is basically the same as that by which sidebands are produced in an amplitude-modulated transmitter. The output of the frequency changer stage before filtering contains components of:

1. the wanted signal on the original frequency;

2. a local oscillator signal;
3. an upper sideband which is the arithmetical sum of the above two;
4. a lower sideband equal to the difference between the first two.

Either of the sidebands contains all of the information of the original signal modulated in the same way. For instance, if the wanted signal was amplitude-modulated with a carrier wave and the sidebands of an audio signal, then each of the newly produced signals (called 'sidebands' in the list above) will also contain a carrier wave and the sidebands for an amplitude-modulated audio signal. The frequency of the carrier of one of the new signals will be the difference between the local oscillator and the original signal's carrier frequency. The other new signal will be at a frequency equal to the sum of the frequencies of the original signal and local oscillator.

For a numerical example, take the case of a 1 megahertz carrier wave, amplitude-modulated by a 1 kilohertz tone, and being received by a set with the local oscillator tuned to 1.5 megahertz. This list shows all the frequencies represented. Before filtering, the output of the mixer stage will contain frequencies at:

499 kHz	
500 kHz	'difference' signal,
501 kHz	AM carrier and sidebands
999 kHz	
1000 kHz	the original signal
1001 kHz	
1500 kHz	the local oscillator
2499 kHz	
2500 kHz	'sum' signal,
2501 kHz	AM carrier and sidebands

The filter must reject all except the one sideband program which appears within its passband. Tuning of the receiver to a particular radiofrequency channel is achieved by adjusting the frequency of the local oscillator. If the receiver has only one frequency changer stage, the frequency selected by the filter will always be the difference between the wanted radio frequency and the local oscillator.

Several different names may be used in connection with the frequency-changing process. The circuitry may be described as a *mixer* or a *convertor*. The process may be called *mixing*, *translation*, *conversion* or *heterodyning*. The fixed frequency of the filter is usually called *intermediate frequency* which is commonly abbreviated to IF. The group of components which selects and amplifies the output of the mixer to produce a signal for the detector is commonly called an *IF amplifier*.

In a high-performance receiver, the IF amplifier may contain several stages of amplification, the filter, and possibly several other tuned circuits as well as some other signal-processing functions.

9.4 IMAGE FREQUENCIES—PRESELECTION

Each time a receiver which uses a frequency changer is tuned to a particular channel, it is also potentially sensitive to another channel as well. In the example in Section 9.3 where a local oscillator on 1.5 MHz is used to tune a signal on 1.0 MHz, the same half-megahertz difference is also produced by a signal on 2.0 MHz. The name given to this other unwanted frequency is *image frequency*.

Once the image is converted to the IF, there is no way it can be separated from the wanted signal. To prevent trouble from this cause, a receiver must have selectivity built in before the frequency conversion. The degree of selectivity required is nowhere near what is required for filtering of adjacent channel interference, but it must be sufficient to block out reception of that frequency.

For equipment designed for local reception of strong signals on the MF broadcasting band, a single tuned circuit which can serve the double duty of input coupling transformer may be all that is required.

At the other end of the performance scale, when a receiver is required to resolve a weak signal in the presence of much stronger ones, and in cases where the wanted frequency is such that the intermediate frequency is only a small proportion of it, then the adequate suppression of image responses requires much higher selectivity and may be one of the more difficult specifications for the designer to achieve. In the case of a general coverage receiver designed to be tunable from less than 1 MHz to about 30 MHz with an IF set at 455 kHz, adequate preselection may require three or four tuned circuits, all must be tunable over the full range and all must be made to track the tuning of the local oscillator accurately as it is changed.

9.5 THE DIODE DETECTOR

Because the original modulating signal is not directly a component of the transmitted radio wave, the receiver must include a suitable detector circuit to recover it. There are probably as many types of detector circuits around as there are brands of washing powders, and as with washing powders, most can be fitted into four or five broad classifications.

For amplitude modulation, the simplest detector contains only a diode rectifier, a capacitor and a load as shown in the circuit in Figure 9.9. With the correct choice of components, the diode will allow current to flow at one peak only (could be either the positive or negative peak) of each radiofrequency cycle and the capacitor will hold its charge until the next peak comes along. The resulting voltage across the capacitor will be a good copy of the original signal at the input to the transmitter.

9.6 AUTOMATIC GAIN CONTROL (AGC)

Signals presented to the input of a high-performance receiver may vary over a very wide range. The weakest usable signal may be only a fraction of a microvolt, and the strongest a fair proportion of a volt. The output is best kept at a particular level. To reduce changes, the DC component of the detector output is filtered and fed back to earlier amplifier stages as a control signal to adjust their gain.

In a high-performance voice communications receiver, this facility may be capable of limiting variations to the extent that a 60 dB range of input levels is reduced to an output variation of only 2 or 3 dB.

The type of AGC circuit used must be matched to the modulation to be received. For single sideband reception, a wrongly designed AGC function could very largely remove the intentional level variations due to modulation making the signal almost inaudible. For reception of frequency and phase modulation, AGC is not used at all. The limiter circuit in the IF amplifier performs a related function, and any effect the AGC circuit would have would be to reduce the efficiency of limiting.

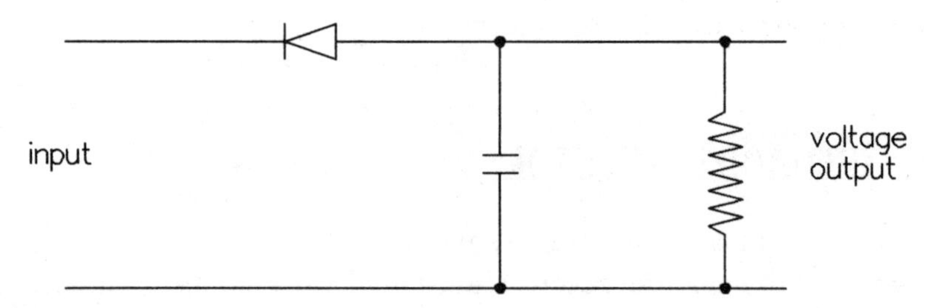

Fig. 9.9 *A diode detector*

9.7 DETECTING OTHER FORMS OF MODULATION

The diode detector only works well with amplitude modulation. For modulation of other types, detectors working on different principles are needed.

For single sideband reception, a frequency translation circuit may be used to effectively shift the carrier to zero frequency; the sideband is then exposed as the original modulating signal. This process is called *product detection* or *synchronous detection*. It may occasionally also be used with amplitude modulation or television signals.

Frequency-modulated signals may be detected by a combination of a limiter and discriminator. The limiter is a tuned amplifier arranged so that it is normally overloaded; the output is a more or less constant amplitude signal. The principle of the discriminator is that two tuned circuits are used, and a diode AM detector is connected to each one. The circuits are tuned slightly off the carrier frequency—one in each direction. The difference between the outputs of the two diode detectors is a good copy of the original signal at the transmitter input.

An FM detector using a different principle is the *ratio detector*. Phase change due to the relative tuning of a tuned circuit is sensed by a phase detector to expose the original modulation. Ratio detectors have an inherent rejection of amplitude variations and so can be used in non-critical situations without a separate limiter, or in higher performance equipment with less severe limiting than is required with a discriminator.

Note that the circuit described as a limiter in relation to FM detection is not the same as the equipment used for preventing overmodulation of a high-fidelity broadcast transmission. The principle is more like a very severe form of the clipping described in Section 4.13.

The vast majority of detectors in off-the-shelf receivers can be identified in principle as either diodes, product detectors, limiter/discriminators or ratio detectors. More complex demodulator circuits can usually be shown to employ the principles of one of these four circuits.

9.8 MULTIPLEXING

There are many cases where a single radio system is required to carry more than one channel of information. The most common examples are the addition of stereo information to FM broadcasting and the use of microwave bearers to carry up to several thousand telephone conversations between exchanges.

One workable method involves treating the input to the transmitter as a 'spectrum' of frequencies similar to the relevant part of the radio spectrum. Each channel of information can then be modulated onto a separate carrier frequency; the whole lot combined is then treated as one signal to modulate onto the radiofrequency carrier for transmission. At the receiver, the demodulation into individual channels is done in two or more stages to unravel the modulation of the transmitter. This process is called *frequency division multiplexing* (FDM). The combination of channels modulated onto *subcarriers* is called a *baseband* and the system that transmits a baseband from place to place is called a *broadband bearer*. The modulation processes do not all have to be of the same type; for instance, in a broadband telephony system, the channels are modulated onto subcarriers by single sideband and then the baseband is frequency-modulated onto the bearer.

For digital signals, a different form of multiplexing is possible. The broadband bearer can carry a stream of data signals at a very high rate but they do not all have to come from the same source. Synchronised switching of transmitter and receiver to different channels can be used to give a digital bearer multichannel capability. This process is called *time division multiplexing* (TDM). TDM can be the most efficient way of carrying multichannel digital signals, and modern practice is that where traffic requires a combination of voice and data, the voice signals are digitised and adapted to the protocol of a TDM broadband system.

9.9 NOISE LIMITING AND ERROR CORRECTION

For all radio systems, the limit of sensitivity is set by how well the received signal overcomes and can be separated from noise and interference. A weak signal can be amplified to any required amount but if the signal cannot be separated from noise, then the noise will also be amplified by the same amount.

The first step in separating signals from noise relates to the selectivity of the receiver. Most sources of noise are broadly distributed over the spectrum so a receiver that is accurately tuned and restricted in passband to only the required channel will offer a better final signal-to-noise ratio than a broader band unit.

From then on, the more you can specify about what will be recognised as an intelligent message, the better able you are to discard that which is not.

Interference from distant lightning is usually short pulses of high intensity; if you can reasonably predict the level of the incoming signal, you may use a clipper to reduce the energy of all pulses over a certain level.

If the noise is generated by equipment drawing power from the electrical mains, it is very likely to be related in frequency to the mains (either 50 or 60 Hz) and may be rejected with a notch filter, or limitation of the low-frequency bandwidth of the output.

In the case of television, it is known that all the energy of the signal is at frequencies very close to harmonics of the line frequency (15.625 kHz for the 625 line PAL system). There are times when a device called a *comb filter* will reduce the energy of signals that are not harmonics of the line rate. The same comb filtering technique can be used with a stream of digital data and there are also some other special provisions such as error correction available for data signals.

For data that is a program or other sequential set of instructions or that includes money values etc., the corruption of one bit of the whole message will destroy the value of all the transmission. There are error-correction techniques which involve adding some extra bits, to each word and block of data, which can signal when a word or block contains an error. This information can then be used to trigger a retransmission of that section as often as is needed until it is received error-free. This extra information may be called either *parity bits* or *checksums*.

An extension of this principle uses enough extra bits in a special form of coding so that the position of errors in the bit stream can be identified and corrected. Procedures of this type are known as *forward error correction*.

In severely noisy conditions, the system can be arranged so that data is sent in very short segments which are immediately transmitted back again to the source and matched with the input. If the returned signal does not match exactly, the original is retransmitted until a match is achieved.

The error-correction system used for a particular service must be matched to the expected error rate. The most powerful systems in terms of ability to correct corrupted data use the most transmitting capacity and can slow the rate of data transfer even in times of clear conditions.

SUMMARY

Most receivers can be shown to have circuit components giving the function of each of the blocks in Figure 9.10. The blocks in the diagram are complete circuit functions such as amplifiers or filters each of which contains a multiplicity of components. For a few very simple circuits, some of the blocks may not appear; for instance, a crystal set, which is the simplest of all receivers, contains only the preselector, detector and output transducer blocks. At the other end of the scale, a high-performance communications receiver may have more than one frequency conversion and a separate IF amplifier for each step of the conversion; a general-coverage receiver may have several different detectors for different modes of modulation and several output buffers for connection to any of a range of output devices.

The vast majority of receivers you will encounter will have at least some components to provide the function of each of these blocks.

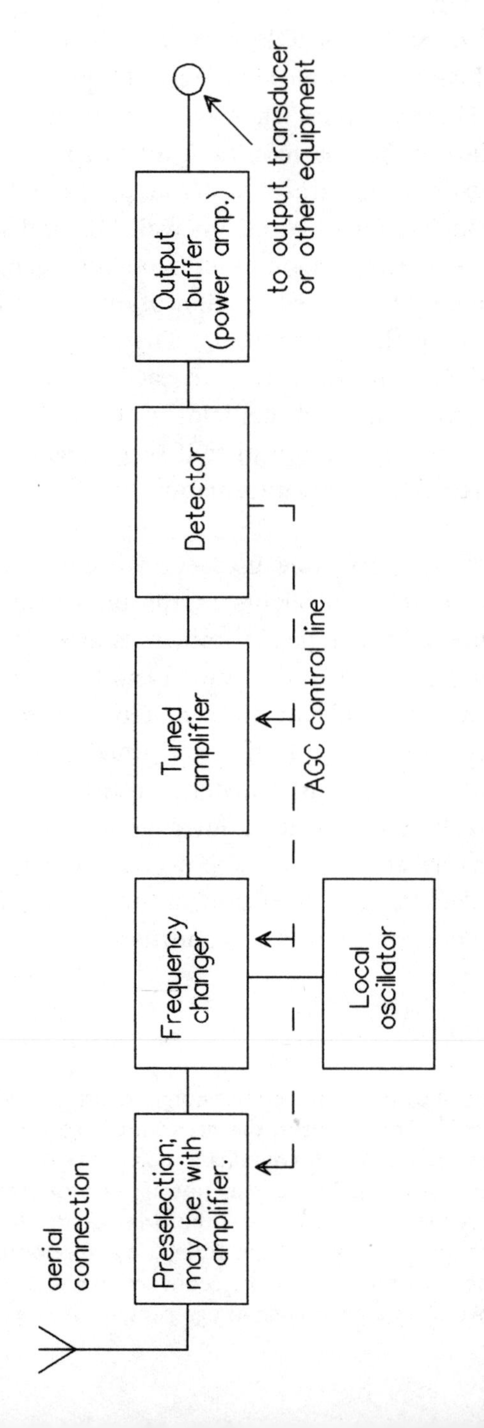

Fig. 9.10 ***A block schematic diagram of a typical receiver***

CHAPTER 10

A VISIT TO THE ZOO: ELECTRONS AND OTHER STRANGE BEASTIES

This chapter contains a number of short sections which are not necessarily connected. The sections of this chapter are detailed explanations of topics mentioned in passing in other parts of the text.
If you are reading through this book for interest as if it were a novel, you will not lose the thread of the story if you simply note the existence of these topics for your reference when needed and go on to Chapter 11.

10.1 MATTER, COMPOUNDS AND MOLECULES

All of us and all of our possessions, the ground we walk on, the food we eat, the air we breathe and all of the physical world that we can see, smell or touch is described by the word *matter*. It is the general term used to describe all physical substances in all their forms. Matter is made up of *molecules* and *atoms*.

A *compound* is a substance made by chemical combination of atoms of at least two different types. A molecule is a group of atoms chemically bound together.

A single molecule of a compound is the smallest quantity of matter that has the chemical properties of that compound. Different compounds have different-sized molecules and a particular molecule may be as small as two atoms (most of the gases in the atmosphere consists of molecules with only two atoms) or as big as a screwdriver handle! Those very highly impact-resistant plastics that are used for screwdrivers, chisels and tools of that type are made of long chains of small chemical units based on the carbon atom. In a high-impact plastic, the chains are so completely cross-linked that it is suggested that the whole moulding becomes one huge molecule.

Probably the largest molecules that can be said to be truly indivisible in the chemical sense are the long chains of proteins in your own body which are many thousands of atoms long but are still of submicroscopic size.

Most of the matter that we deal with in our daily lives is fairly tame in the sense that we do not see great release of energy simply because we walk on some of it or move lumps of it around. The energy is there, however, and in some parts of the universe such as the interiors of stars, the matter is being ripped apart and the raw energy exposed.

The reason that matter can be so tame and well behaved in the presence of so much energy is that for the most part, all the forces that would release the energy are opposed by an exactly equal force with the result that normally everything is held in its place. The chemical changes that we do see releasing energy (fire for instance) are the result of slight momentary imbalances in these forces that cause a rearrangement and set up a new balance.

10.2 ATOMS—ELEMENTS

The study of chemistry is all to do with elements and compounds and the reactions between them. It can be a fascinating study but it is not the major subject of this book. However, the definitions of elements and compounds, and the distinction between them, is relevant. As explained in Section 10.1, compounds are chemical combinations of atoms of different types. An *element* is a lump of matter in which all of the atoms are of the same type. Many millions of different compounds are possible but there are only 92 naturally occurring elements. At room temperature they may come as gases such as oxygen, nitrogen, hydrogen or neon, as liquids such as mercury or bromine, or as solids such as iron, copper, aluminium, sulphur or carbon.

Each element has its own characteristic type of atoms. The major components of atoms are:

1. *protons*: positively charged 'heavy' particles;
2. *neutrons*: 'heavy' particles with no electric charge;
3. *electrons*: light, negatively charged particles.

In a stable neutral atom, the protons and neutrons are all bound together in a *nucleus* at the centre, and the electrons move around at various distances from the nucleus in accordance with a fairly complex set of laws. An atom can be compared in layout to a solar system with the nucleus taking the place of the Sun and the electrons in place of the planets. The analogy runs out very quickly when you look at details, but as a starting point, it gives us a rough idea.

The type of element represented by a particular atom is determined by the number of protons in the nucleus. The elements form a series called the *Periodic Table* based on the number of protons. The first few members of this series are:

1 proton = hydrogen, a gas with 2 atoms per molecule
2 protons = helium, an inert gas with 1 atom per molecule
3 protons = lithium, a metal similar to sodium

If the number of protons in the atom changes, then the atom becomes a different element. This is an atomic nuclear reaction and is usually accompanied by release or absorption of vast amounts of energy.

Of the 92 natural elements, most can be divided on the basis of broad similarities of chemical properties into one of two classes: *metals* or *non-metals*. There are a few that form a small group of semiconductors with properties in between those of metals and non-metals.

Sizes and density of atoms are of interest. Atoms are extremely small—several hundred times smaller than the wavelength of visible light—so they will forever be invisible to the eye no matter how much magnification could be applied to them. A yardstick called the *angstrom unit* is often used to measure the diameter of atoms. The angstrom unit is one part in 10 million of the length of a millimetre. That is, $\frac{1}{10}$ of a nanometre (or 10^{-10} metres). Atoms range in size from about 0.6 to 3 angstrom units. Despite their small size, atoms are mostly empty space! Almost all of the mass is concentrated into the incredibly tiny nucleus. The actual particles of the atom are almost inconceivably dense. It has been calculated that if all the space could be squeezed out of a group of atoms, the resulting 'plasma' would be so dense that one cubic centimetre (that is, about $\frac{1}{10}$ of a matchbox-full) of it would have a mass of 100 million tonnes. Material such as this could not exist on Earth as we know it, but is thought to be present deep in the interior of large stars which have burnt out all their nuclear fuel, then cooled and collapsed under gravitational forces.

10.3 ELECTRONS

This section is written on the assumption that you have read and understood Sections 10.1 and 10.2.

We humans do not really know a great deal about electrons. They are so small and fast that they are very hard to apply measurements to. We can measure other bigger things by comparing them to an electron, but to build up a portrait of an electron itself requires an even smaller bit of matter to compare it with. What follows are some of the basic facts we do know about electrons rather than a complete portrait.

1. An electron carries a unit negative electric charge.
2. An electron has a mass of 1/1837 of a proton.
3. Electrons are affected by magnetic fields. If an electron is moving across the lines of force, it will be deflected at right angles to both its own movement and the lines of force and follow a circular path. If it is travelling in close to the same direction as a line of force, it will have a spiral path which tends to follow that line.
4. Gravity on a terrestrial scale has almost no effect on electrons. (There are some parts of the universe where gravity may affect the movement of electrons, but nice people don't go there!)
5. Each electron is attracted to positive electric charges in its vicinity. The exact nature of this force is not known but some facts about its operation can be fairly well defined:
   (a) The resultant force is the vector sum of all the forces applied by all the relevant positive charges, with the inverse square law applied to distances.
   (b) There is no evidence of any sort of quantising being applied to the forces. In theory, every positive charge anywhere in the universe has an effect on every electron; in practice, only the nearest few ever need to be considered.
   (c) The resultant force is resisted only by the inertia of the electron and deflections due to magnetic forces.

In practice, electrons are in constant motion. For most of them the motion is orbital around a particular nucleus and for some materials this orbital motion accounts for almost all the electrons.

Each electron carries energy. When it is close to its parent atom it travels very fast and energy is mainly kinetic; when it is further away, kinetic energy is traded for potential energy. The total of the two forms of energy remains constant and the principles of quantum mechanics do apply to the energy of an electron. Electrons can gain or lose energy only by absorption or radiation of an electromagnetic wave, and when that happens the energy change is in discrete packets called *quanta*. Electrons in particular positions in an atom can absorb or radiate quanta of a particular size only, and this fact can explain all of the observable characteristics of an electron in association with an atom. The quantum theory does, however, lead to some fairly complex calculations in its explanation of electron behaviour.

The point to remember about quanta is that they apply to the energy carried by an electron, not to the forces acting on it.

10.4 CONDUCTORS AND INSULATORS

To understand this section you will need to have read Sections 10.1, 10.2 and 10.3.
There are some materials in which one or two of the outer electrons from each atom are free to leave the atom and wander off through the bulk of the material. Overall there are the same total number of electrons as protons, so the material is electrically neutral, but the free electrons cannot be assigned to any particular atom. In other types of materials, almost all the electrons are tightly bound to a particular atom or molecule.

Electrical conductivity depends on these free electrons. The best insulating materials are those where all the electrons stay assigned to a particular atom. In good conductors the material can be thought of as being like a gas or soup of electrons.

Among the chemical elements, the best conductors are all metals, and those with highest insulation resistance are non-metals. Among compounds, there is a class of mineral salts called *electrolytes* which dissolve in water and render the solution a very good conductor of electricity.

Some of the most effective insulators are compounds which fall into the class of *organic* chemicals. Another class of chemicals that form good insulators are the ceramics which are oxides of metals—the oxidation process ties up the free electrons into a particular molecule.

For all chemical substances, electrical conductivity is raised when the substance is brought into a field of ionising radiation (X-rays and gamma rays).

A vacuum is a special case—its conductivity depends on the surface conditions of electrodes within the vacuum chamber. If the surface of the negative electrode is heated, or subjected to ionising radiation, or formed into a very sharp point so that electrons can be freed, then the vacuum itself offers no resistance to current flow. In most other cases, a vacuum is one of the best possible insulators.

10.5 IONISATION

When an atom is electrically neutral, the number of electrons in orbit (in the planet positions) is exactly the same as the number of protons in the nucleus. Sometimes a neutral atom can either lose or gain one or more of the orbiting electrons and when that happens, the whole atom acts as an electrically charged particle. Atoms that have lost an electron have a net positive charge and are attracted to points which are negatively charged. Conversely, atoms that gain an electron are particles which have a negative charge. These atoms with an electric charge are called *ions*. The process that forms them is called *ionisation*.

When a chemical compound is formed between one or two atoms of a metal combining with one or two atoms of a non-metal, the chemical bond is formed by the metals losing an electron and becoming positive ions and the non-metals taking up those electrons and becoming negative ions. From a chemist's point of view, that statement is a brutal oversimplification of the process of a chemical change. If you want to know more about the details of the process, the topic to look under is 'Acids, bases and salts' in the general subject of inorganic chemistry.

There are other ways ions may be formed. If you make a lump of matter hot enough, the energy added will be enough to cause some electrons to be knocked out of their orbits. That happens to all forms of matter but the temperature required is different for different substances. Highly energetic radiation (X-rays, gamma rays, cosmic rays etc.) will change the energy level of an electron and may remove it from a particular atom.

Negative ions which are free to move will follow the same path as a free electron but will travel a lot more slowly. Positive ions will move in the opposite direction; for instance, positive (metallic) ions are attracted to the negative electrode in an electroplating bath.

10.6 THE WATER ANALOGY

There are a lot of similarities between electrons in a conductive wire and water in a pipe. In both cases there is a flow resulting from pressure and the rate of flow depends on both the pressure and the physical features of the pipe or wire (diameter, length and type of material). (See Fig. 10.1.)

Volts and amperes ('amps' for short) can be equated very closely to plumbing units for water pressure (pounds per square inch or grams per square metre) and flow rate (gallons per minute or litres per second etc.).

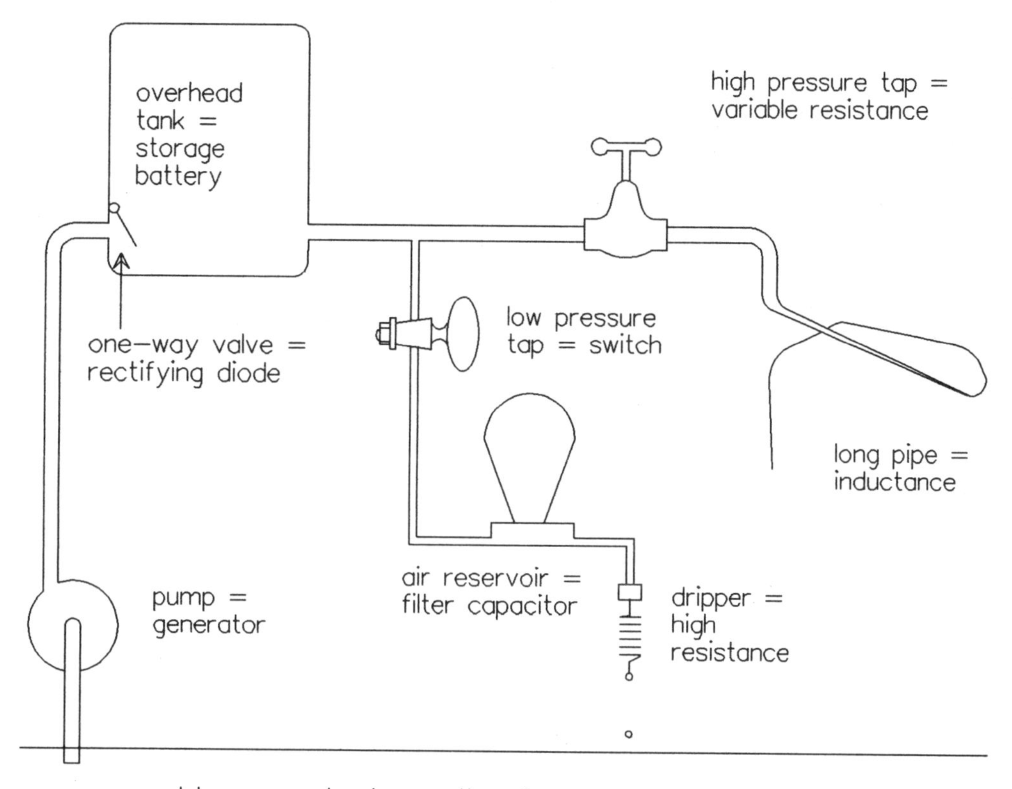

Fig. 10.1 *Analogy between flow of water in a water system and flow of electrons in an electrical circuit*

In a water system the flow is usually controlled by a tap. There is, however, an intrinsic property of resistance for each particular installation which sets a limit on the maximum possible flow. This intrinsic resistance depends on factors such as the length and diameter of pipes, the number of bends and the roughness of the inner surfaces. If measured it would be expressed in a form such as 'litres per second per kilogram per square metre'.

In an electrical circuit, control is most often achieved by switching so that the current is normally controlled by the built-in resistance of the circuit. In a similar fashion to the water pipes, each electrical circuit has an intrinsic property of resistance which depends on circuit geometry and the material it is made of. Taps in the water pipe can be equated with variable resistors.

The effect of inductance can be roughly equated with the inertia of the water in a long pipe; the simile is not exact but fairly close.

Capacitance can be equated with elasticity in either of two ways. The air reservoir on a hydraulic ram has exactly the same effect as the filter capacitors of a power supply. A diaphragm in a sealed box totally filled with water with a pipe connection on each side of the box equates with an in-line capacitor. Water flowing in one side stretches the diaphragm and displaces water out the other side, but not the same water. Continuous flow through the box is not possible.

The water analogy runs out when electromagnetic radiation is considered. There is no mechanism for radiation of energy when the water flow is changed. The nearest similarity is the water hammer effect when sound is radiated if a flow of water in a long pipe is suddenly stopped. Electromagnetic radiation is, however, a vastly different process from the sound of a water hammer.

10.7 BIG AND SMALL NUMBERS—PREFIXES

The range of numbers involved in measurements on radio systems can be very large. Frequencies of interest can be from a few hertz to thousands of millions of hertz. Voltages may range from millionths of a volt up to many thousands of volts. Currents, resistances and reactances may show a similar wide range of values of interest. Path losses and amplifier gains expressed as power ratios may show even wider ranges. (Power is voltage multiplied by current.)

When the metric system was devised, the basic units were named and there were names given in prefix form to the powers of 10 values of these units. As time has gone on, the scientific community has adopted these prefixes for general use with a range of measurements other than the original metric system.

The full 'dictionary' of prefixes for every power of 10 is too complicated for ordinary users to remember and so only every third one corresponding to powers of 1000 have remained in common use.

The prefixes relevant to 'big' numbers are:

kilo-	for 1000 times (10^3)
mega-	for 1 million times (10^6)
giga-	for 1000 million times (10^9)

For small numbers the relevant prefixes are:

milli-	for one-thousandth part of (10^{-3})
micro-	for one millionth part of (10^{-6})
nano-	for one thousand millionth part of (10^{-9})
pico-	for one million millionth part (10^{-12})

Sometimes it is possible for calculations to be transferred in scale; for instance, calculations of volts, ohms and amps can equally be applied to volts, kilohms and milliamps. However, there can be traps; for instance, power calculated in *watts* from volts and amps will be changed to millivolts, milliamps and *microwatts*.

10.8 VOLTS, AMPS AND OHMS

There is an electrical unit called a *coulomb* which is a quantity of electric charge—a particular number of electrons. The number is very large, roughly 6 followed by 18 zeros. The coulomb can be equated with litres or gallons of water—it is a particular-sized bucketful of electrons. If exactly one coulomb passes a particular point of a piece of wire each second, then one *ampere* of current is flowing. Amperes (commonly shortened to amps) measure electrical current flow and, as noted, can be equated very closely with flow rate in a water system.

The *volt* is a unit of electrical pressure. Electrical pressure in volts is needed to drive current in amps through a resistance. The unit of electrical resistance is *ohms*, named after the man who discovered the relationship between pressure, current and resistance. An electrical pressure of one volt applied across the terminals of a resistance of one ohm will cause a current flow of one coulomb per second or one amp.

The relationship between voltage, current and resistance is described by *Ohm's law* which says that the current flowing in an electrical circuit is proportional to the voltage applied and inversely proportional to the resistance in the circuit.

Ohm's law is one of the most basic things you have to learn when you start to study any of the whole group of subjects that come under the title of Electrical Technology.

When electrical pressure in volts drives current in amps through a resistance, power is converted into heat energy. Originally the volt was defined as the potential difference (electrical pressure) between the terminals of a particular type of cell made with certain chemicals. The definition has been changed so that the volt is now a derived unit defined in terms of coulombs, time in seconds and power flow in watts.

10.9 WATTS AND JOULES

Watts and joules are not specifically electrical units. They are the SI (Système International) units for power and energy in all their forms. A *joule* is a particular quantity of energy, like a bucketful. Power in *watts* is a measure of the rate of flow of energy. The names of these units (and most of the names of units in the previous few sections) are those of famous early researchers in modern physics.

The relationship between joules and watts is roughly the same as the relationship between a tip-truck and a conveyor belt. A joule of energy is one watt of power flowing for one second; a watt is a rate of flow (like the conveyor belt) of energy of one joule per second.

By human standards, both units are quite small. A strong man working hard can produce about 160 watts. The standard horsepower is 746 watts, and one watt is about the power of a cat working hard (has anyone ever persuaded a cat to work hard?). The 'unit' that electricity is sold in is the *kilowatt hour* (kWh). There are 3.6 million joules in each kilowatt hour.

At the other end of the scale, the watt is a very large unit compared with the minimum that our senses can detect. Even one milliwatt (note this is talking about milliwatts not watts) of sound power at an audible frequency actually getting inside our ears would be a very loud noise indeed and would probably cause hearing impairment if allowed to continue. Similarly, one milliwatt of visible light actually flowing into the pupil of our eye would be bright enough to cause instant blindness. The limit of sensitivity of our eyes and ears is better than a million million times fainter than that.

The output of a radio transmitter is usually measured in watts because it is a constant power flow which is there for as long as the PTT switch or HT contactor are operated. Certain types of interference, in particular the noise of distant thunderstorms, are commonly measured in joules because in these cases each event is unrelated to any other and the disturbing effect depends on the total energy of each pulse.

10.10 THE DECIBEL SCALE

As mentioned in Section 10.9, our ears and our eyes are extremely sensitive to power inputs within their operating frequency ranges. Both organs also have a self-adjusting mechanism similar in effect to the automatic gain control circuit of a radio receiver so that the apparent range of power variations is reduced. Early experimenters found that the loudness we perceive of sounds is not directly related to the power or energy of them. The drop in level of a steady tone to half the power will not be heard by a listener as a tone that is half as loud; there will in fact be a barely noticeable change in loudness.

To better match the way we hear sounds, a logarithmic scale was developed. (Logarithmic scales are not uncommon when natural factors have to be measured; for example, the pH scale for measuring the acidity of a solution is another logarithmic scale.) For every 10 times increase in the power level, 1 is added to the unit recorded. This unit was called a *bel* after the man who first successfully transmitted a voice over a pair of wires. The bel was soon found to be a rather 'blunt instrument' for daily use and a finer gradation was needed, so the unit of $\frac{1}{10}$ of a bel was soon commonly used and given the name *decibel*.

The decibel scale is a ratio between power levels originally achieved by comparison of sounds; it does not have a fixed reference. The power level meters in an electronic circuit need to be calibrated to a fixed reference, and so a number of variations of the decibel scale have been evolved. The most common are:

dBm = decibels referenced to one milliwatt
dBW = decibels referenced to one watt
dBμV = decibels referenced to one microvolt

These are the measurements that may be found respectively at the audio input to the transmitter (and the line output of a receiver), at the aerial connection of a transmitter, and at the input of a receiver.

Although it is possible to calculate the exact decibel ratio of a pair of power measurements, it is more usual in practice for people who use decibels to memorise a few simple figures, such as:

3 dB is roughly 2 times the power.
10 dB is 10 times the power.
20 dB is 10 times the current or voltage or 100 times the power.

From then on, the scale is used for roughly estimating relationships.

CHAPTER 11

FIRST IN MAINTENANCE

11.1 CHAPTER OUTLINE

Location and rectification of faults in electronic equipment is a huge and complex subject, and for radio systems, the vagaries of propagation add a further range of possible sources of failure. Each item of equipment that is sold today has available a technical service manual and each manual has at least a chapter on test procedures and fault mechanisms specific to that piece of equipment. If all those chapters of fault-finding information were extracted and placed together, those dealing only with equipment that is manufactured specifically to be used for radio communication would amount to a library bigger than the largest encyclopaedia but still would not be a comprehensive list of all possible symptoms and all possible causes.

In modern times the volume of information that a service technician would need to carry to be proficient in fault-finding, even for such a range as all the models of all brands of equipment for a particular class of service, is more than one person can carry and is forcing the servicing industry into a situation where each technician must specialise his/her knowledge to only a few models; ever increasingly the method of operation is that the equipment is removed from service and sent to a central depot where the specialists for that model operate.

This volume of information makes it impossible to give any sort of worthwhile overview of the whole servicing industry in one chapter of a book of general principles such as this. There is, however, one facet of servicing that is worth covering in some detail. This chapter will focus on the situation where an operator of domestic or mobile equipment who has little or no access to electronic test equipment is faced with the task of deciding whether a fault is an electronic failure in equipment that he/she has control of or the fault is in some other part of the system. The information given will be aimed at being able to prove that a particular item or section of the installation has a fault so that that item can be referred to the appropriate service technician with some confidence that bench tests will show the fault. Proving that a fixable fault exists and isolating it to a particular box is sometimes described as *first in maintenance*.

11.2 STEP ONE—DON'T WAIT UNTIL IT IS FAULTY

The process of fault-finding is one of collecting information. The information you seek consists of observations on what is different from normal operation, and in that respect, the more you know about what is normal for your particular piece of equipment, the more accurate and definite you can be about what has changed when the fault occurs. The actual observations you make will depend on what the piece of equipment is and what its purpose is, but the following list details some of the types of observations that may be valuable later:

1. For mains-powered items, is there a dial light, pilot light or other display that is on all the time the item is plugged into a working power point?
2. For battery-powered items, is there a 'low battery' indicator?
3. Does the loudspeaker blurt if you turn the power on with the volume control fully up? Does it blurt if the volume control is at zero?
4. What does the receiver sound like when it is tuned off a station?
5. How does mobile equipment perform when you take it to the very edge of its working range? What happens when you take it a bit further away?
6. For equipment that includes a microprocessor controller, there may be an initialisation procedure at first switch-on which has several separate steps with different indications at each step. Observe closely the first few seconds of operation and memorise what displays are normal at each step.
7. Do the settings of any of the controls vary appreciably in the first 20 minutes or so of operation after you first switch on from cold?
8. Does the performance of the item vary when the weather is particularly hot or particularly cold or when it is raining?
9. Is there a 'standby' mode of operation? What indications do you observe when the item is in the standby mode?
10. Can the equipment be switched for multiple functions? Are there switch positions for functions for which your particular model is not equipped?

This list is by no means comprehensive but the general principle is that the more familiar you are with normal operation (particularly in relation to switch-on procedures), the more chance you will have of identifying an abnormal condition.

11.3 WHEN IT DOESN'T WORK—PRELIMINARY CHECKS

When you first become aware of an unexpected response, look and think before you touch. Compare what you observe with what you remember about what is normal. Check that there are cables plugged in where you expect them to be and that all controls are in sensible positions. Note check *all* controls; if something has been knocked by accident it will not necessarily be obvious where the correct position should be.

For the usual small items of electronic equipment where transistors or integrated circuits and their associated components are mounted on a printed wiring assembly and supplied with power at a stable low voltage, there are parts of the circuit where the failure rate can be extremely low. When these units are handling low-level signals, the most common faults (refer to Section 11.10) are in those components that have mechanical movement such as switches and adjustable controls, plugs and sockets, motors or loudspeakers. The relatively non-technical procedure of wriggling everything that moves while watching for a correct output is often a sensible next step in fault location. However, there is a proviso to that; it is possible to upset adjustments which were correct and which may then require a complex and expensive procedure to reset. Adjustments of this type usually have a small dob of a sealing compound added after the adjustment is set correctly. As a general rule, do not move any of these sealed adjustments.

For circuits that handle higher power levels, the presence of heat causes acceleration of aging and very often the first effect of fault conditions is to make components overheat. It is worthwhile for the operator of these types of equipment to get to know where the normal hot spots are and to memorise how warm 'normal' is. The appearance of any new hot spots or changes to normal temperatures are signals of the early stages of faults.

If you are someone whose fault-finding experience is related mainly to mechanical devices, it will be well to remember that by comparison, electronic assemblies are more affected by high temperatures, corrosion, vibration and very quick pulses on the battery or power line, and less affected by dust (except for the components with mechanical action mentioned above), than would be expected for comparable items of machinery.

11.4 HOW THE PROFESSIONALS DO IT

There is nothing worse for a service technician than to be presented with a piece of equipment with a fault report that makes no sense, and then to put it on the bench and do a full set of proof of performance tests only to find it is working perfectly.

Unfortunately, that general scenario happens so often that many of the big service organisations are geared to it as the normal process. They have developed quick and efficient ways of duplicating the original manufacturer's final quality control tests on the assumption that they can return the equipment to original specifications and prove that they have. In a lot of cases they even go as far as undervaluing or ignoring the user's report of the fault.

When a fault is indicated, there are often checklists for that make and model based on this sequence: 'If symptoms are that, that and that, replace this, this and this component and it should work.'

Signal tracing and actually finding the fault are often slow and expensive ways of working. For the bulk of faults which can be traced to a particular box, a standard, often automated, series of tests will clear the fault even if it does mean replacing half a dozen components of which you can be sure that all except one of them was perfectly all right.

The other thing that professional organisations do that is relevant from the operator's point of view is that they make a distinction between a quick look at the overall system aimed at restoring service by changing plug-in modules or other large sections of the system, and the more detailed work of component location and replacement. The quick overall look and restoring of service by changing to standby equipment may also be called 'first in maintenance'.

11.5 PARTICULAR VALUE OF OPERATOR'S REPORTS

Intermittent faults are always a potential problem for service technicians. In general, nobody can fix a fault that he or she cannot observe, and the major hurdle is to cause the fault to occur on the bench. If you know that a piece of equipment has a fault which may be intermittent, then that fact is the most important piece of information to pass on. Add a note to the box with the word 'INTERMITTENT' written in big black or red letters.

The next most important information is all you know about conditions that will bring on the fault: temperature, humidity and power supply voltage are always relevant; and whether you can cause interruption by mechanical means—pressure in particular places, vibration, moving of cables or holding the box at a particular angle are also points to test.

In circuits that include a microprocessor, there may be some trip functions that behave like intermittent faults but in fact the condition is there all the time just waiting for the combination of factors which triggers the trip. The operating conditions that existed immediately before the failure are the ones to record and the indications on any screen display will lead the technician to the trip function that has operated.

11.6 CORRECT INFORMATION SAVES MONEY

The person who knows they are lost and knows they don't have a clue about where they are is only half as lost as the person who is quite sure of their exact location—but isn't there! Similarly, the operator who is quite sure of which box is faulty but has in fact picked on the wrong one will make the whole job more expensive, and the more energetic they are in investigating the fault, the more expensive the final cost will become. It cannot be stressed too strongly that giving no facts at all is better and less expensive than giving wrong information.

Consider the following sequence: An organisation servicing two-way radio systems receives a mobile with selcall number 259 to fix. They check it on the bench and find no fault. A few days later, mobile number 907 appears and also checks fault-free. Then mobile number 712 turns up, is found to have a fault, is fixed and is returned. Next week, number 552 and number 895 are brought in; one is faulty and is fixed and one is checked and found clear.

The servicing organisation would either need to have a particularly efficient fault-history-recording method, or a technician who has a personal interest in that company, to find the common factor, but at $50 to $70 for each proof-of-performance test, this fault at the base has already cost a couple of hundred dollars which could have been saved by an operator realising that all sets with a selcall digit '9' are not responding.

The more information you can collect about a fault, the cheaper the fixing will be, *provided that* the information you collect leads in the right direction.

11.7 LOOK AT THE WHOLE SYSTEM

There are three major reasons why equipment tested on the bench may show the 'all-clear':

1. The fault may be intermittent and only show up under certain environmental conditions.
2. There may be a difference between operating conditions in the field and manufacturer's test conditions which happens to be important to this piece of equipment.
3. The person who removed the equipment from service made a wrong guess about the location of the fault.

In all the fault-finding work I have done over the years I have never yet found a case where a paying customer reported a fault when absolutely nothing was wrong. What I *have* found is that the actual location of the fault is often very different from where the operator thinks it is.

If you, as an operator, are aware of a fault condition and wish to localise it, *the most essential* step of your process is for you to be aware of the whole system and not to remove any part from suspicion unless you can positively prove it is clear—which usually means proving the fault is in some other part.

Remember also that your (and other operators') degree of understanding of the system and the way you operate it are both themselves factors in the system for fault-finding purposes. For radio services in particular, the environmental factors and the understanding of operators may be particularly relevant. When the system is designed to provide communication to a multitude of randomly selected locations, it almost goes without saying that no one person will be able to run an eye over all parts of the system and trace signals along looking for anomalies. Consider, for instance, the typical block diagrams in Figure 11.1 in Section 11.8 on the next page.

11.8 BLOCK DIAGRAMS

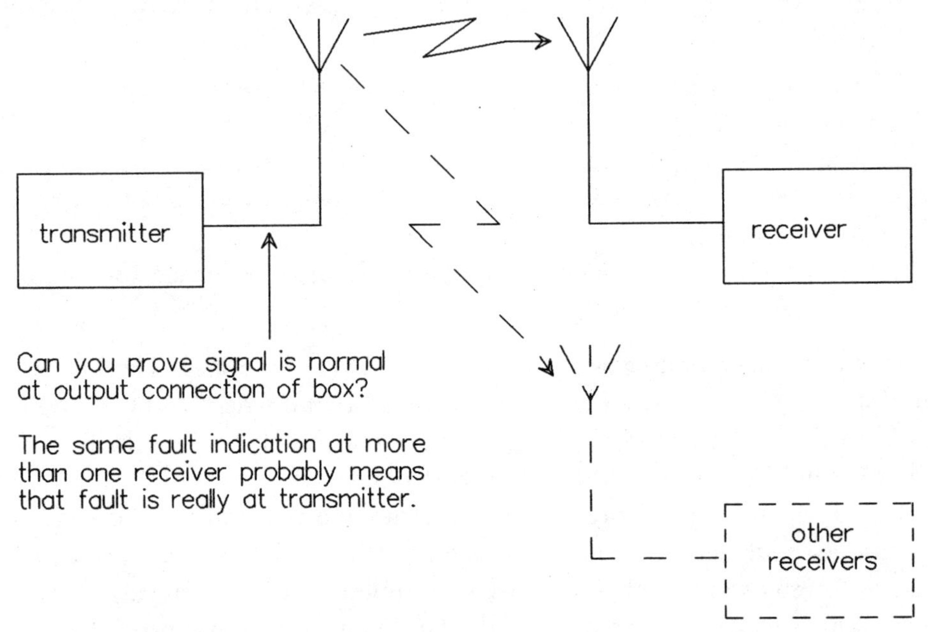

Fig. 11.1(a) *Locating faults*

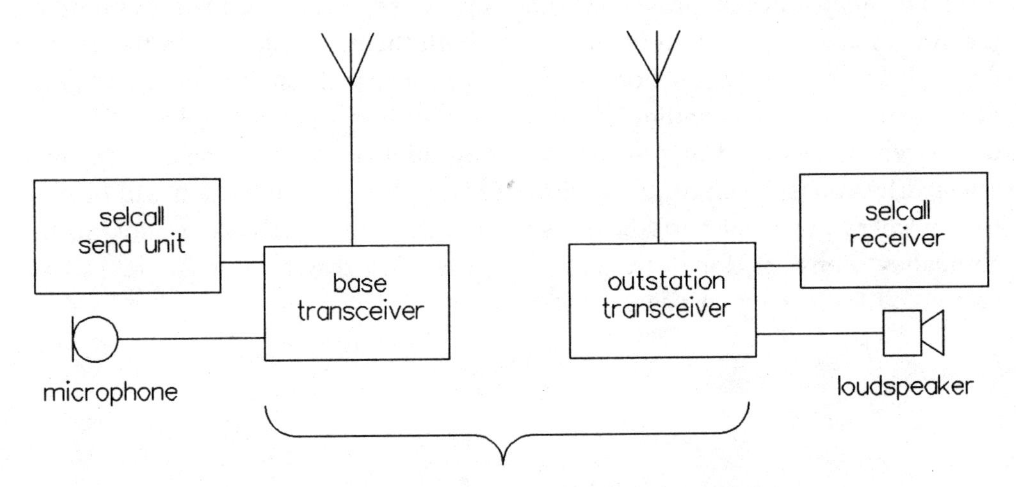

Fig. 11.1(b)

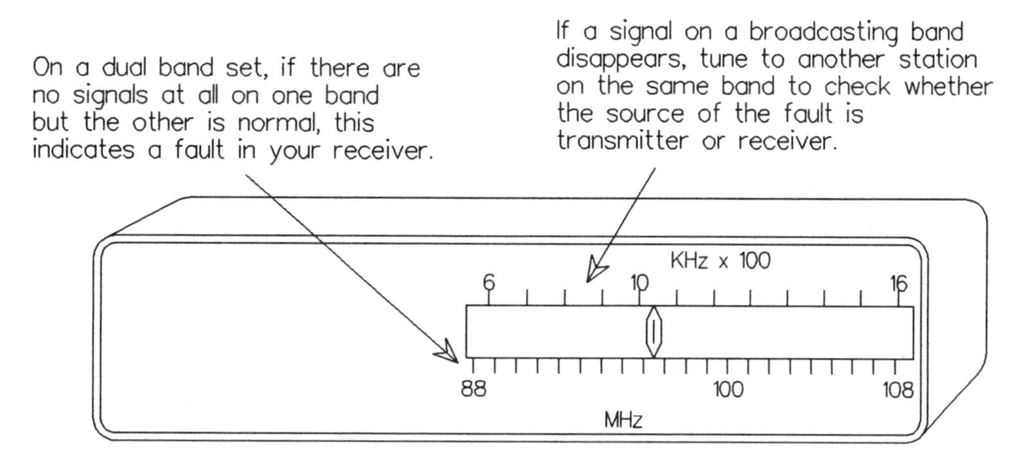

Fig. 11.1(c)

11.9 FAULT LOCALISATION

In a few cases the location of the fault will be made obvious by mechanical damage. For instance, if you are the operator of a mobile two-way radio system and you come to work after a stormy night to find bits of antenna elements scattered all over the yard, the location of the fault is defined and you take the appropriate action to restore that function before you bother about any more detailed tests.

As an operator you will probably waste resources overall if you attempt to do the job of the service technician. The technician's expertise is in knowing the internal functions of each piece of equipment. After a short time to gain familiarity with your system, you will probably find you are more aware of the overall functions and operating conditions of your particular system than is a technician who has an intimate knowledge of the boxes that compose it. Your aim should be to collect information that will localise the fault and lead a technician to a particular box or assembly.

Step 1: Do not wait until a fault shows up.
Use the times of normal operation to learn as much as you can about your system and understand what is normal. Refer to Section 11.2 for more detailed suggestions.

Step 2: When a fault occurs, collect all the information you can about what is abnormal. In particular, if you know the exact time of failure, write down whether there was any abnormal event such as a power surge or lightning strike at the same moment.

Step 3: Above all, *do not* give wrong or misleading interpretations to the information you do collect. Do not be too judgmental at first about whether the information you collect and record is relevant or not, but do concentrate on ensuring that what you do record is accurate.

Step 4: After all the information is collected, have a think yourself about what it might all mean and see if it gives you a suggestion as to the location of the fault. Do not discard any of the recorded information at this stage.

For radio services there is always a major division in the system. The first question to ask of the information is: 'Can you prove that either the transmitter or receiver is fault-free?' In most cases, a piece of equipment can be proved by substitution. This will imply different tests for different circumstances of course. For instance, if a broadcast band receiver is being used on a strong local station and it suddenly goes quiet, you may suspect that the fault is with your receiver but you can quickly check it by tuning to other channels. If other stations sound normal, you can expect that the fault is in the transmitter.

On a two-way network with one base and many outstations, you may lose contact with one station but be able to call others. This may not be as definitive as the first example, although, if you find you have lost contact with more than one outstation, you should suspect some function of the base. It is extremely unlikely that two outstations would develop faults with the same characteristics at the same time.

If the system consists of one transmitter and one receiver, or if you are the operator of an outstation with access to only one piece of equipment, the decision on which to suspect may become more difficult. In many cases it will be best to start with the receiver; if you can listen to or somehow display the output of the detector you may be able to define a general 'liveliness' that is an indication of a receiver working at full sensitivity with no input. This is an example of a case where knowing what is normal can be very valuable.

If the equipment you have access to operates on multiple channels and you can prove that it is normal on any one channel, then that is a very valuable piece of information. Although it may not isolate any one piece of equipment, it will remove about 90% or more of the whole system from suspicion and will direct the service technician to a small group of components in each box.

For telemetry and data links, you may find that no one can carry out any sensible fault location without some fairly sophisticated test equipment. In these cases, all you can do is prove that a normal signal is going into the link and that it is not coming out the other end. The actual tests you do and measurements you notice will depend on the type of service. For instance, listening to an HF transceiver while you remove the aerial connection is a fairly definitive test for the whole receiving section and the aerial because on the HF band you can expect that there will be more noise generated by lightning than by the front-end circuits of the receiver. The same test on a UHF set, however, would be inconclusive and, for a cellular telephone system, would definitely be undesirable because it would leave the transmitter running at full power with no load on the output! The more you know about what is normal for your system, the more definite you can be about the location of faults when they occur.

11.10 SPECIFIC TESTS FOR PARTICULAR TYPES OF SERVICE

This section gives some simple tests that can be performed without electronic test equipment. You may find them valuable in the particular services mentioned but you also may be able to adapt them in principle to other related services.

For broadcast receivers

If a signal is received but is distorted, slightly adjust the tuning to check it is centred on the channel. If the signal is still distorted, check for overloading if the equipment is very close to a local transmitter (try tuning to a more distant transmitter) or, if equipment is battery-operated, check for a flat battery. Most forms of distortion other than these are electronic faults that will require the attention of a technician.

One notable exception to the above is if you are operating a broadcast receiver in the HF (international short-wave) bands and distortion is associated with fading—this is selective fading due to multipath propagation of the signal. If the same program is available on a higher frequency channel, then this other channel should give clearer reception.

If the signal is weak and affected by noise then different causes can be indicated at different times. If the receiver is a car radio and the noise is associated with the car's ignition or alternator, this may indicate a wrong connection of the power source from the vehicle battery. Generally all other types of impulsive noise are being received as radiated signals and the only protection you can get from them is to change location to a spot that is closer to the transmitter and further from the source of the noise. If the receiver is mains-operated, the noise is a buzz or hum and the aerial connection is via a coaxial cable, then a broken earth connection of the coaxial cable may be indicated. There are many thousands of possible sources of noise and the cure for the problem will depend very much on just what the noise is; for example, noise associated with a weak signal often indicates a defective aerial connection.

To test for the path of entry of the noise signal, disconnect the aerial. If the noise is still present, it either originates from the power supply or is being picked up by induction into the speaker leads. If the noise disappears when the aerial is disconnected, it is being received as a radiated signal and the source is outside of the equipment.

Note that for sensitive AM and SSB receivers and for all FM systems, there is an internally generated noise which sounds like a rushing wind. This noise is normal and simply indicates that the receiver is operating at full sensitivity with no signal input.

If the signal suddenly disappears during a program, try tuning to another station on the same band. If the signals from other stations are normal, it means the transmitter has gone off air. If no other signals are heard, it means your receiver has failed. If this is the case, can you prove power is still getting to the receiver? Are pilot lights, dial lights or other indicators still on? If the receiver is part of a multifunctional audio system, do other functions work normally? If not, check for disconnected speaker cables. If some functions work but others do not, this almost certainly indicates an internal fault which will require a technician's attention.

If the receiver has previously been working but now cannot be switched on or, at switch-on, briefly flashes and then is dead, check for fuses on the equipment itself. Also, if power is by mains supply, check, using a portable lamp, that power is actually available at the mains socket.

For mobile phones

For most types of faults that cause interruption of service, switch the unit off, wait 30 seconds and then switch on again. Watch closely the switch-on procedure. If the 'log on' message is displayed, then almost all of the electronic circuits of your set are proven correct.

If there is no indication of life at all, check for faulty power connections, in-line fuses in vehicle installations, or dirty battery contacts in hand-held units.

If the set is a hand-held unit and successfully logs on but after a few seconds or a minute or so reverts to a fault condition, then this may indicate a battery that has lost its charging capacity. The battery will need to be replaced.

If the set indicates a weak signal, check your location. In hilly areas or where there are multistorey buildings, the UHF signals used can be very patchy, or signals in the local environment may exhibit a fine-grained (within a few centimetres) pattern of reinforcement and cancellation.

If your phone is installed in a vehicle and the phone suffers from noise which appears to be associated with vehicle functions (ignition etc.), check the power connection. For maximum noise immunity, there should be a pair of wires which carry both positive and negative connections directly from the set to the vehicle battery and are connected as closely as possible to the terminal posts of the battery.

Note carefully: Do not under any circumstances disconnect the aerial or antenna connection from a mobile phone that is switched on. The system operates in the duplex mode which means that the transmitter is delivering power to the aerial all the time a call in either direction is being set up or is in progress.

If the fault is a system fault in either the base or the central controller, it will appear to clear itself within a few hours, so if your set fails to log on, it may be worth trying again later in the day. If failure to log on lasts for more than a couple of hours, it is almost certainly due to an internal fault in your set that will require the attention of a service technician.

For a personal mobile radio (PMR) system (two-way radio) which operates on a direct channel in the VHF or UHF bands and uses frequency modulation. (This includes UHF CB operating on the simplex channels.)

For vehicle-mounted sets, if there is no sign of life such as pilot lights or illuminated display or LCD panels visible when you first switch on, check for faults in the power connections and check also for fuses mounted on the rear panel of the transceiver itself. For hand-held transceivers, this may not be a definitive test; 'power on' indicators waste battery power so there may be no indication of 'power on' apart from receiver noise from the speaker or earphone.

Operate the set with the receiver mute (may also be called 'squelch') open. You should be able to hear the rushing-wind noise that is characteristic of a sensitive receiver operating with no input. If you hear no noise, there is either an interruption to the power supply (check for flat battery or dirty contacts on hand-held units or check connections and fuses with vehicle-mounted equipment) or an internal fault in the receiver which will require the attention of a technician.

If you happen to receive a signal during the course of the above test which is able to silence the rushing-wind noise of the receiver, then this is a good indication that the aerial and the receiver are all functioning normally.

If your system uses selcall and you can contact your base, ask the operator to send a selcall to your mobile and then make a voice announcement while you listen with the mute open. If you hear the selcall tones and the voice you recognise but it has no effect on the alarm system of your station, this is a sure indication of a fault in the signalling system which will require a technician's attention.

If either you hear no signal or the signal is so weak that it does not silence the rushing-wind noise, move the mobile station very close (within shouting distance) to the base and try again. If operation is normal at short range but not at ranges greater than half a kilometre or so, the probable cause is an open circuit or short circuit in either the aerial, its base or the feeder cable to it. Faults of this type can be positively identified with an electrical ohmmeter if one is available.

If your mobile uses a 'gain whip' antenna, and range for normal operation is reduced but not as severely as described above (maximum operating range perhaps 2 to 5 kilometres), try replacing this element with another properly tuned whip as a first step. If that test cures the problem, discard the faulty whip; if not, you will need to refer the fault to a technician.

If you can prove that the aerial and receiver are working correctly, try to transmit a call at a time when you know someone is listening. If your call is not heard, the fault is almost certainly either in the electronics of the transceiver or its associated microphone. If the microphone is a plug-in unit and an identical spare is available, test it by substitution. If the fault is proved to be in the transceiver electronics, refer it to a technician.

For VHF/UHF transceivers operating on repeater-based systems. (Also relevant to UHF CB operating on 'duplex' channels.)
All of the tests detailed for the directly operating PMR systems mentioned above are relevant except for the 'close range' tests of aerial function. For a repeater-based system, even if the two stations are operated side by side, the signal does a round trip of many kilometres to the repeater and back.

With repeater systems, traffic density on the channel is normally higher than for a small direct system so if you listen for a couple of minutes with the receiver mute open and hear nothing except rushing-wind noise, you can be fairly sure that there is a fault either in your aerial or at the repeater.

If your mobile is one of a fleet, check with other mobile operators. If any others work normally, the fault is in your aerial; if they all experience the same problem, the fault is at the repeater.

For HF land mobile transceivers (Flying Doctor type). (Part of this may also be relevant to CB transceivers operating on the 27 MHz band.)
These are normally mounted in a vehicle with a pair of heavy-duty wires making a direct connection for power to the vehicle battery. If there is no sign of life

when you first switch on, check for open-circuit connections in this power cable and check also for fuses on the rear panel of the transceiver.

Note that if you intend to rely on your transceiver for any life support or emergency service function, it is sensible to run through the following series of checks each time you start a journey before you leave your home base.

Operate the receiver with the mute open and the volume control well advanced. If there is no traffic on the channel at the time, you should hear a noise which is partly a steady rushing-wind noise and partly an impulsive noise due to distant thunderstorms. Remove the aerial connection from the transceiver (normally a coaxial cable)—the noise volume should decrease and the impulsive portion of it disappear. If there is no change in noise output from the receiver, the aerial is either wrongly tuned for the transceiver channel selected (check taps on multitapped whips etc.), or faulty, or its feeder cable is faulty. In many cases faulty connections on aerials and cables can be positively identified with an electrical ohmmeter if one is available.

If in the above test there is very little or no noise from the receiver even with the volume control flat out, the indication is for a fault in the internal circuits of the receiver. If your transceiver is operated by a remote control head and uses an extension speaker, wriggle all the plugs, sockets and cables associated with that section of it. Any noise at all in the speaker in response to mechanical movement will give an indication of the fault location. If the receiver gives no noise output and the fault is not due to a contact or cable defect that you can fix yourself, refer it to a technician.

Transmit a test signal which includes periods of silence as well as speech. If there is a transmitter output indication (these days usually an LED but could also be a meter), observe that it is extinguished when you are silent and glows brightly flashing in step with your speech.

Transmit a very soft whistle which you gradually make louder. The transmitter indicator should show some glow even with very low levels of audio input, and should glow at full brilliance when your whistle is still significantly softer than normal speech.

If there is a continuous output at times when the vehicle's engine is off and you are transmitting with no speech, this possibly indicates a parasitic signal being radiated which may cause interference to other services. If you observe that indication, do not make any more transmissions and take the whole installation to a technical specialist. There are some causes of parasitics which are electronic faults in the transceiver but there are also some which can be caused by the layout of the installation within the vehicle so you will need to give the technician access to the whole vehicle.

If the above test is clear when the engine is off but there is an indication of transmitter output when the engine is running, this indicates electrical noise being induced on the power feed line from the vehicle battery to the transceiver.

Complete lack of indication of transmitter power output even when you speak will be due either to an electronic fault in the transceiver or to a fault in the microphone. If the microphone is a plug-in unit and an exact replacement is available, test it by substitution. If the fault is proved to be in the transceiver, refer it to a technician.

If the transceiver has a 'power on' indicator as well as a 'transmitter output', and you notice that the 'power on' indicator also varies as you speak, the indication is that either the vehicle battery is flat or there is a high-resistance connection in the power cable to the transceiver. The 'high' resistance may still be too small to be read with a normal ohmmeter but the power dissipation is such that you may be able to locate it by the heating effect of the current. Make a transmission of a steady whistle for about 10 seconds and then feel along the cable—the change in temperature will only be slight but should be noticeable by comparison with unheated portions of the cable. The location of the warmest spot will lead you to the location of the fault.

If your mobile station passes all the above tests, choose a channel which predictions indicate should give strong clear signals to a local base. (Section 5.8 gives information on how prediction of HF radio signals works.) If you are familiar with the system and the time of day is such that you would expect the base to be attended, transmit a test call and ask for a signal report. If you are new to the system listen to the receiver for perhaps an hour or so while you work around the vehicle; this will give you an understanding of the way experienced operators use the system. At a time when you think the channel is clear and the base operator is listening, transmit a test call and ask for a signal report.

If you hear signals but they are weak, maybe with some fading, or if the signal report indicates your transmission is only being received at low strength, switch to a higher frequency channel and repeat the calling procedure. If you hear nothing at all from the expected base but you can hear other signals that sound as if they may originate from further away, select a lower frequency channel and repeat the calling procedure.

The position of the microphone in relation to your mouth can be important. If the signal report is that your signal is strong but distorted, try again with the microphone held further away (up to about 150 mm) and to the side of your mouth. If your transmission is still distorted, this indicates an internal fault in the transceiver which must be referred to a technician.

When equipment is used in marine or tropical environments

Tests on transceivers for marine use are comparable with those relevant to land-based types. For VHF marine sets you can follow the same test procedure as for VHF/UHF PMR systems, and HF marine sets can be directly compared with those used in association with the Flying Doctor network.

The particular significance of marine and tropical environments is due to the effect of corrosion which can cause latent faults particularly on plug-and-socket contacts which have no effect until the joint is moved. In these cases, prevention is better than cure and it is recommended that you follow a regular schedule of joint inspection and contact cleaning so that these latent faults are exposed and corrected before they cause failures of the equipment in service.

CHAPTER 12

THE HUMAN FACTORS

12.1 THE CHAIN FROM BRAIN TO BRAIN

In the final analysis, the performance achieved by a radio system is only partly determined by the excellence or otherwise of the equipment. The purpose of all radio is the transfer of information, and radio systems are chains of pieces of equipment linked together in certain ways. At each end of the chain there is a human brain, and unless the message leaving one brain forms a sensible and recognisable concept in another human brain, the purpose of the system has failed.

The early experimenters in radio had a grand dream to drive them on which was to give everyone in the world the power to transfer concepts undistorted and instantly from, for example, your brain (whoever you are and wherever you happen to be) into another person's brain. The dream was impossible but it wasn't until the equipment was made and being used that the limitations of it were found out.

One obvious limitation was channel capacity. There are at present 5000 million people on Earth and, at a maximum, about 100 000 million hertz of channel space. To give each person access to even a 2 kilohertz channel would require that all frequencies be reused at least 100 times over.

There is, however, a more subtle but much more restrictive limitation. Each of us is thinking all the time and we need time to digest our own thoughts and even more time to digest the thoughts of others. In the modern world, there is an abundant supply of people who wish to make us understand the way they feel about things, but there is a limitation on how much information each of us can absorb. We are bombarded with raw information all day and every day, with messages falling over each other demanding our attention. The result of all this is not perfect understanding of anything but great oceans of undigested information which threaten to drown us.

12.2 TRANSFERRING UNDERSTANDING, NOT JUST INFORMATION

The modern telecommunications industries have been of most value to two sorts of people: those who seek to sell us something and thereby gain the resources for a loud, strident message demanding our attention, and those who have an interest in collecting information about us to give them power to control us which once again gives them the economic power to buy channel space for their purposes.

For most ordinary people, advanced telecommunications facilities are of little value unless we happen to be employed in one of the industries that uses them. Note that in general terms we all benefit from the increases of efficiency gained by industrial, manufacturing and distribution innovation. The above statement is talking about opportunities for individual private use of 'high-tech' equipment. For each of us there are really only a few people we wish to pass concepts to or from on any regular basis, and in a surprising number of cases, we adjust our lives so that most of those people are the ones we can talk to face to face.

Despite all the technically advanced radio and telecommunications equipment now available, the best and surest way to pass a concept undistorted from one brain to another is still two people standing face to face in a reasonably quiet environment and with something to draw pictures on. A stick on a patch of bare ground is close to as good as you will get. The advantage of a pencil on a piece of paper is storage of the information—the receiver can carry it away and study it more at a later time. When you look at all of our communications systems as complete chains for the transfer of a concept from one brain to another, none of them has given us any clearer transfer of understanding than that simple situation. What electronic technology does offer are enhancements in other directions such as longer range or faster access, but we should always be aware that advantages in these directions are at the expense of trading some of the completeness of transfer of concepts.

12.3 TRADE-OFFS ON OFFER

When compared with two people talking face to face, radio systems offer us advantages in other directions at the expense of trading some of the clarity and completeness of transfer of understanding of the face-to-face situation.

A broadcasting system can, for instance, transfer information to many thousands of other brains at the same time but at the expense of denying the speaker any chance to gauge how the message is being received. This, coupled with the fact that in almost all cases the listeners are volunteers whose attention must be maintained, puts severe limitations on the types of subjects that can be broadcast and the depth of detail that can be gone into without causing the listener to become bored.

At the other end of the scale, a point-to-point link using HF transceivers can extend the distance between two speakers and give them a chance to gauge each other's interest and understanding but at the expense of such severe limitation of audio response that most of the timbre of voices is lost and the communication is reduced almost to the technical precision of machines talking to each other with none of the inflections in voices that convey emotional responses.

In another direction, text and even drawings can be transmitted almost instantly to the other side of the world. However, the drawing must be a finished work before it can be transmitted; it is very difficult to display it as it is being drawn while at the same time explaining the development of the concept.

Thus in all these cases the completeness of transfer of understanding is traded for an advantage in some other direction. For the mass media, the trade is for numbers of listeners; for communications services in general, the offer is distance—more particularly in terms of time saving and cost saving. If the degree of reduction of understanding can be regarded as the price of distance capability or numbers of listeners, then the technical equipment sets that price; the operator has little direct control over it except to choose a different type of service. Whoever is selecting equipment to use for communication should be aware that, for all services, limitations exist and the job of selection is to minimise the practical impact of those technical limitations on the particular type of use being planned.

12.4 'CORRECT' PROFESSIONAL PROCEDURE

Professional operators have over the years found that they can increase the understanding content of the information they send, and operators using professional procedures can increase remarkably the intelligibility of a message during periods of marginal communication by agreeing among themselves to use standardised forms of messages.

It should be made quite clear at this point that there is no one operating procedure which is 'correct' thereby making all others wrong. There is a correct procedure for the Safety of Life at Sea (SOLAS) network which is internationally known and used. There is a correct procedure for communication with aircraft which is not the same as the SOLAS procedure but is similar in a lot of ways. Each of the emergency services has its own procedure which reflects the relative importance of various tasks. Bushfire brigades, for instance, rarely talk about fires in multistorey buildings; metropolitan brigades often do not need to consider crown fires or wind shifts in gullies.

Although these various 'correct' procedures are different in detail, there is a similarity about them which stems from the similarity of their basic aims, that is, whatever message is received will be clearly understood no matter how bad the background noise or how distorted the audio signal.

The basic tool of the professional operators is that they have agreed among themselves to limit the number of possible meanings for each message and to transmit the message in a defined sequence to further limit the number of possible meanings for each section. For ships and aircraft, for instance, normal messages contain information about course, speed, altitude (for aircraft), time of arrival at way points and ETA at the next destination. Other sorts of information are very rare so these messages are stylised in great detail. On the other hand, a bushfire brigade would have little use for most of that information but it would need to report the level of water in the tank and the expected number of minutes before the tank will be dry and relief required at their operating location. The 'correct' operating procedure for each class of service reflects these differences in requirements.

There are a few procedures which are almost universal. A phonetic alphabet based on the English language has been developed which is used by all disciplines. It consists of 26 words which have been chosen so that the sound and shape of each under conditions of severe distortion is as diverse as possible from the other 25. There is a standard way of pronouncing numbers, also chosen with the aim of maximising diversity for distorted reception. Call signs are used to identify a particular station no matter who is using the microphone and there are a few 'pro-words' such as 'over', 'standby' or 'out' which are almost universally recognised.

12.5 FOR MORE INFORMATION

This book is an entry into several quite diverse fields of interest. This final section will give some outline information about each of those possibilities and where to look for more, but without going into too much detail about any one of them. This, in a sense, is the parting of the ways—the meaty details of each option will in most cases be specific to a particular industry or type of hobby, and a greatly detailed treatment of any of them would be tedious and boring to any readers not directly interested in that aspect of the subject.

Your decision as to what direction to take will be related to whether your interest is for radio as a hobby or as part of your job. There will also be a choice depending on whether you wish to learn the technicalities of the electronic equipment functions, or whether you are content to treat all that as black boxes and concentrate on operating procedures which leads you to the study of ergonomics, applied psychology and other related subjects.

Let's look first at the hobby aspects of the subject. You may, for instance, have started reading for general interest just to keep up with the basics of a number of branches of technology. You may have found that all you needed to know at this stage was in the first four or five chapters of this book. Keep it on your bookshelf—it is intended to be a book you can read in sections and it is almost certain that now you know that the information is available you will find a need to refer to it from time to time.

The electronics associated with radio have as much as possible been ignored. If that subject interests you at the hobby level, there are two books that will give you a good start: the yearly *Handbook* of the American Radio Relay League (ARRL) and the *Handbook* of the Radio Society of Great Britain (RSGB). Both are intended to give all the technical information that a person interested in amateur radio would need to pass the licence examination and assemble and put on air a good quality station in the amateur class of service. Both start with absolutely basic electronics, so they are good for a general interest in the technical understanding of radio systems of all types.

Libraries carry a range of books on the electronic aspects of radio under the classification number 621.3815. Their collections include professional designers' handbooks which in a sense start where the ARRL and RSGB handbooks leave off, dealing with of electronics and mathematics at a greater depth.

Within the broad field of radio technicalities there are a number of more specialised branches on which information can be found in libraries under the following classifications:

Microwaves	621.3813
Communications satellites	621.3825
Radiocommunications and antennas	621.384
Television	621.388

If you wish your hobby to be more than just armchair exploration, you can choose to be a listener/reporter, a transmitter/experimenter or a tinkerer with equipment. In each case, you will find that other people who share your interest have formed themselves into clubs and associations to promote their own branch of the general hobby of 'radio'.

A very wide diversity of special interests is possible. There are, for instance, people who listen for overseas broadcasts of the international short-wave stations and send signal reports to them; they are in regular contact with people all over the world. In another direction, there are people with a hobby interest in propagation research who are recognised as being leaders in the field and whose knowledge on their particular aspect of the art is respected by the full-time professionals in universities and other research institutions. And for the tinkerers, there is a myriad of circuit functions which are only occasionally required, and so will never be mass-produced and marketed, but each nonetheless has its own niche as a one-off item. From the point of view of hobby interests, there is an almost never-ending range of specialities that come under the broad umbrella of the radio art.

Your interest in radio may be because it is part of your job or you may be looking at it as a possible future career. In that case the range of interests in the marketplace is almost as broad as the range of possible hobbies.

You may be working as an operator of a radio system and feel that your work has been hampered by a lack of understanding of what the system is actually doing. In that case your interest is probably highly specific to the particular equipment you are operating and the information you seek will probably come from the manufacturers or suppliers of the equipment, either as operator's manuals or marketing information. Marketing information, however, is very often 'a lot of feathers and very little meat'. If you have specific questions about particular operating procedures or anomalies you have observed and you get a chance to ask a technical person about them, you may get good answers.

However, beware of the curse of specialisation. There are people who know absolutely everything you could think of to ask about the electronics of a particular black box and nothing at all about propagation of radio signals. If you observe an effect that is actually due to propagation and ask such a person about it, he or she will at first try to interpret your question in terms of the electronics they know. If you try to run a very big program on a computer with insufficient memory you are given a 'memory full' or 'out of memory' message. The radio art in all facets of every speciality is a huge subject that has a similar effect on the human brain. The procedure that most people use to defend against this factor is to focus their attention on one small speciality. These people who focus on a speciality in this way actually know less about other aspects of the subject than others who are less specialised in their interest. The trick is to choose the right specialist to ask.

There are a number of separate and distinct career paths in 'radio' open to people. This book may, for instance, prompt you to enter into a career in the scientific study of radiophysics. In Australia, most of that work is carried out in the universities in several major centres and at the CSIRO, a Commonwealth Government organisation that has a centre for radiophysics in an outer suburb of Sydney. Using the same field of information as the radiophysicists but in a more applied form, you could have a special interest in field strength surveying. Neither of these avenues has any direct reference to electronics, although in practice, to get entry into them, you would probably be required by the training course to study at least basic electronics.

Electronic servicing can be fascinating work and would have all the mystery and romance of a good detective story but for the fact that it is electrons you are chasing not people. This is technical work involving component manipulation which can range from the fine precision of surface-mounted components on multilayer printed circuit boards up to the scale of crowbars, cranes and forklifts for the heavier components of high-power transmitters.

With the present state of the art, it is probably best not to enter electronic servicing with the aim of making it a lifetime career. The trend in automation of manufacturing is towards making circuit cards with techniques similar to the printing industry, and these days, no one ever repairs books by patching in new pages to replace defective ones. In the very early days of radio, the manufacturer made resistors by shaping them from lead pencils, and diode rectifiers from a selected lump of rock and a fine piece of wire called a 'cat's whisker'. It is now normal for many thousands of resistors, diodes and other components to be built into a single integrated circuit and replaced as a single component whenever any one item fails. The trend in servicing is for ever larger modules and subassemblies to be regarded as throw-away components.

A technician's course of instruction and some experience gained in a servicing job can be a valuable opening into one of the fields that is expanding. One of those is reliability engineering. Another is design and prototyping of circuits for low-volume applications. The manufacturers using automated techniques must have a large-volume market, and that means only certain types of functions can be economically supplied. Their dream is for all consumers to want the same thing. The real needs of consumers will continue to be individual and that means there will be a host of demands for particular items, each of which has only a low-volume market.

The starting point for all jobs of this type is usually a course of instruction leading to a technician's certificate either as a full-time student at a technical college or sometimes by correspondence. There are also some opportunities offered by governments and large commercial organisations for cadet or trainee positions where training is on the job, but the long-term result leads to careers of the same type.

You may have decided that you are interested in radio system operation. This can lead you to such diverse situations as a ship's radio operator on an Antarctic icebreaker, a weather station or lighthouse on a tropical island, or a broadcasting station based in a multistorey building in a large city. It may involve you in the space race or put you at the nerve centre of operations of an emergency service directly concerned with matters of life and death, or it may be that your radio operation is only a minor part of a more general communications duty in which you use telephones and typewriters as well as face-to-face contact for message handling. The operator normally does not need to be an expert in either electronics or the general field of radiophysics, but must know something of each of these in association with the people-oriented skills of ergonomics and psychology.

In a few particular disciplines, the message-handling process is highly regimented and defined but there are still very many cases where effective communication is an imprecise art and even the experts have not yet defined the real basics of it in repeatable scientific terms. One of the problems is that in research on communication skills, we must measure ourselves, and the measurements each of us takes is from the inside of ourselves and from the outside of all other people.

Scientific research on radiophysics, technical work on electronics, the people-oriented skills of system operation, or the hobby interest of just wanting to know how the message gets there, are as diverse as any four subjects could be, but all are taken in under the umbrella of 'radio art'.

If you have read this book and I have succeeded in my aim of transferring understanding of the concepts to you, then you will have taken a first very important step up the ladder that leads to all of them. Where you go from here depends on your interest; at this point the way up branches into several ladders and there is a whole world to explore in each direction. I wish you success in your exploration.

APPENDIX

FOR AUSTRALIAN READERS

A.1 EDUCATION/TRAINING FOR CAREERS IN 'RADIO'

Research in radiophysics: Training starts with a Bachelor of Science degree with radiophysics taken as a post-graduate specialty.

System designer (or a similar engineering profession): Start with a degree in Electronic Engineering and specialise in telecommunications.

Full-time, part-time or correspondence courses, leading to graduate qualifications, are offered by universities. The normal entry path is through one of these offices:

Northern Territory University
PO Box 40106
CASUARINA NT 0811
Phone: 1800 061 963

For intending student applications, the address is:
Student Administration
Attn Admissions
Northern Territory University
DARWIN NT 0909

Queensland Tertiary Admissions Centre
PO Box 1331
MILTON Qld 4064
Phone: (07) 3368 1166

South Australian Tertiary Admissions Centre
230 North Terrace
ADELAIDE SA 5000
Phone: (08) 8223 5233

University of Tasmania
Student Recruitment and Course Information
GPO Box 252C
HOBART Tas. 7001
Phone: (03) 6226 2514

Universities Admissions Centre
(Initial contact must be on a form obtainable from all New South Wales universities.)
Locked Bag 500
LIDCOMBE NSW 2141
Phone: (02) 9330 7200

Victorian Tertiary Admissions Centre
40 Park Street
SOUTH MELBOURNE Vic. 3205
Phone: (03) 9690 7977

WA Tertiary Institutions Service Centre
39 Fairway
NEDLANDS WA 6009
Phone: (08) 9347 8000

Hands-on work on equipment is normally carried out by people who are described as 'tradespersons', 'technicians' or 'technical officers'. Training for civilians in these careers usually involves a course at a TAFE college. Courses offered range from a Certificate course in Radio Servicing to an Associate Diploma in Electronic Engineering. Courses are assembled from training modules which are standardised Australia-wide.

A large proportion of the people who undertake TAFE courses work as apprentices or trainees. There are many hundreds of organisations which offer apprenticeships and traineeships, ranging from government departments such as the Defence Forces, through semigovernment corporations and large commercial companies, to some quite small businesses providing specialist services in particular fields. You may also undertake a TAFE course as a full-time student with no ties to an employer provided that places are available in the course.

TAFE colleges are listed in the alphabetical telephone directory under the word 'Employment'. (Employment, Training and Further Education—Department for). There is nearly a page of references in each book.

The Defence Forces also have a substantial commitment to training in the general field of electronic telecommunications and telemetry, of which radio is a part, and practically every branch of the radio art is represented somewhere in one or more of the three Forces. All careers in the Defence Forces have a common starting point. The postal address for all initial contacts is:

Reply Paid 2600
Australian Defence Force Careers
GPO Box XYZ
(state capital city) Postcode is 2001, 3001,
4001, 5001, 6001 etc. as appropriate.
Phone (Australia-wide): 13 1901

The Recruiting Centre has a booklet which details the career options available and the location of the training centre that deals with each specialty with the intention that you should work out what path you are most interested in following.

In addition to the TAFE colleges and Defence Forces, the following commercial training organisations are known to offer certificate courses by correspondence in technical subjects associated with radio and telecommunications:

Royal Melbourne Institute of Technology
Office for prospective students
GPO Box 2476V
MELBOURNE Vic. 3001
Phone: (03) 9660 4570

Offering all TAFE level subjects.

International Correspondence School (Australasia)
1 Waltham Street
ARTARMON NSW 2064
Phone: (02) 9201 4511

Relevant courses offered include 'TV and Video Repair' and 'Journalism'.

Stotts Correspondence College
140 Flinders Street
MELBOURNE Vic. 3000
Phone: (03) 9654 6211

Relevant courses include 'Basic Electronics', 'Radio Servicing', 'Television Servicing' and 'Radio Receivers'.

There are many training and promotional organisations which periodically offer exhibitions, conferences, seminars or workshops for time periods ranging from one day to a week or more. In most cases these are intended for those already working in the industry as extension or refresher courses.

A.2 FOR TRAINING OF OPERATORS OF PARTICULAR SYSTEMS

Ship station operators

As at 1996, there is a major change in progress to maritime safety and rescue procedures. A Global Maritime Distress Safety System (GMDSS) is being phased in to be fully operational by 1998, and this will include the traditional radiocommunication facilities in an integrated network of all the ship's communication functions. Emergency communications under the GMDSS are controlled by:

The Australian Maritime Safety Authority
PO Box 1180
BELCONNEN ACT 2616
Phone: (06) 279 5000

Training courses in GMDSS operation (not necessarily only for ship's officers) are conducted by:

The Australian Maritime College
PO Box 986
LAUNCESTON Tas. 7248
Phone: (03) 6335 4711

Cooloola Sunshine Institute of TAFE
PO Box 130
BUDDINA Qld 4575
Phone: (074) 446 152
(Subject to accreditation in early 1997)

Department of Manufacturing, Mechanical and Marine Technology
Attn Marine Section
Royal Melbourne Institute of Technology
GPO Box 2476V
MELBOURNE Vic. 3001
Phone: (03) 9660 4410

Marine Technology Centre
Sydney Institute of Technology
Level 7 Building W
827–839 George Street
BROADWAY NSW 2007
Phone: (02) 9217 3402

South Metropolitan College of TAFE
Business Centre
41 South Terrace
FREMANTLE WA 6160
Phone: (08) 9239 8111

Due to the introduction of GMDSS in the maritime industry, it is not now possible to get a job on an Australian ship specifically as a ship's radio officer; the operation of communications and distress-calling facilities is an integral part of the training of ship's deck officers. All training for both officers and other ranks starts with a course of Integrated Ratings Training, conducted by the Australian Maritime College (see above for address and telephone number). Entry to the College is gained either by appointment with one of the ship-owning companies or through:

Maritime Industry Training Committee
40 Beach Street
PORT MELBOURNE Vic. 3207
Phone: (03) 9646 2059

The GMDSS applies only to large ocean-going ships or to those that have a high value or high risk due to the number of passengers or the particular type of cargo carried. For all ships equipped with marine satellite communications or capable of the digital selective calling techniques used in the GMDSS, the operator must hold a GMDSS General Operator's Certificate of Proficiency. That qualification is issued and administered by the Australian Maritime Safety Authority.

For small ships (for instance the fishing fleet), the appropriate qualification is a Restricted Radiotelephone Operator's Certificate of Proficiency (RROCP), which is normally gained from an examination conducted by the Spectrum Management Agency or one of their appointed delegates. Training for this qualification may be either by private study or through a TAFE college or the Coastguard organisation. There are also a number of commercial or private sources of training in general boat operation, coastal navigation and sea safety which include training for the RROCP in their courses.

Inshore boating

There are some voice channels allocated in the 27 MHz range for marine use in a self-regulating manner similar to the citizen's bands. Apart from the restriction that these channels should not be used by land-based operators, they can be used by anyone with the appropriate equipment.

The Spectrum Management Agency can be contacted at either the Central Office or at one of the Area Offices listed below. The address of the Central Office is:

Executive Manager
SMA Customer Services Group
PO Box 78
BELCONNEN ACT 2616
Phone: (06) 256 5555

Area Offices are located in:

Adelaide	Phone: (08) 8228 6860
Albury–Wodonga	Phone: (060) 24 7711
Brisbane	Phone: (07) 3238 6322
Cairns	Phone: (070) 31 4266
Canberra	Phone: (06) 256 5577
Coffs Harbour	Phone: (066) 51 5452
Darwin	Phone: (08) 8941 0366
Hobart	Phone: (03) 6223 5644
Melbourne	Phone: (03) 9685 3555
Newcastle	Phone: (049) 29 6899
Perth	Phone: (08) 9323 1717
Rockhampton	Phone: (079) 22 2312
Sydney	Phone: (02) 9922 9111
Townsville	Phone: (077) 72 2977

Coast stations (international shipping)

Until recently a coast station operator required the qualifications of a seagoing radio operator, but with the advent of GMDSS, a new set of conditions will apply. Initially, GMDSS will automate some functions, so opportunities for new employment will be very limited for a few years. When new staff are taken on, they will be required to hold the Radio-communication General Operator's Certificate of Proficiency and have passed the training course for GMDSS (see list above).

Coast stations are operated by Telstra Mobile Satellite and Radio Services, and the appropriate address for initial contacts seeking general information is:

National Marketing Manager
Telstra Mobile Satellite and Radio Services
79 St Hillier's Road
AUBURN NSW 2144
Phone: (02) 9202 4000

Limited coast stations

In some cases there may be no particular qualification required; in others, the RROCP is appropriate.

Aircraft pilots

Each stage of the pilot's training includes instruction on the operation of the relevant communication equipment, and the licence for each stage of training authorises the pilot to operate the relevant aeronautical radio equipment. There is information about becoming an aircraft pilot available at most attended airfields.

On some large commercial aircraft, there is an officer designated 'Flight Engineer' who is not a pilot but who must be able to operate the radio equipment and so must have the same radio qualifications as a pilot. The title of that qualification is Flight Radio Operator's Telephony Certificate of Proficiency. Information on this is provided in a handbook issued by the Civil Aviation Safety Authority (CASA).

Airport communication and navigation equipment

The most common operators of airport radio-communication equipment are employed as either air traffic controllers or flight service officers. These people are employed and trained by the Civil Aviation Safety Authority which has a representative at every airport that carries regular public transport air services.

Technical operation and servicing of navigation aids at airports is a highly specialised field. The starting point for this work is a qualification as a technician or technical officer (see Section A.1) but before any particular person can do work on a navigation aid, they must be accredited for that particular type of equipment. The responsible body for work on airport equipment is

Airservices Australia
Phone (Australia-wide): 13 3550

There is a field office of Airservices Australia in each state capital. Contact information on each can be found in the alphabetical listing of the metropolitan telephone directory of your state.

Airworthiness certification is carried out by the Civil Aviation Safety Authority. There is an airworthiness inspector in each state capital, and initial approaches should be made to that office. If that is not possible, then the Head Office is at:

GPO Box 2005
CANBERRA CITY ACT 2601
Phone: (06) 222 2111

Airport ground crew

There is a need for people to operate ground-based mobile equipment (for instance fire tenders and aircraft tugs) on the aeronautical network and also for ground crew to operate aircraft radio-communications for testing and adjustment while the craft is on the ground. The Civil Aviation Safety Authority issues a Radio Operator's Certificate of Proficiency for that purpose. The initial contact for people interested in that qualification should be to the CASA office of each state capital city (alphabetical listing of metropolitan telephone books). If that is not possible, the Head Office of the relevant section is:

Personnel Licensing Branch
Civil Aviation Safety Authority
PO Box 219
CIVIC SQUARE ACT 2608
Phone: (06) 222 2111

Base station operators—land mobile

In land-based mobile or outpost networks (such as the Royal Flying Doctor Service (RFDS) bases, fire brigades and other emergency services and bases for commercial use), the base requires a station licence which is held by a particular person or incorporated body. The licensee may appoint others to do the actual on-air operation but it always remains his/her responsibility to ensure that proper operational procedures are followed. Because the station licensee retains responsibility for station conduct, he/she sets the standard for operator's qualifications. No particular operator's qualifications are required by the Spectrum Management Agency and training is usually done on-the-job and is specific to a particular base or system.

Outpost mobiles and similar

Stations must carry a licence but no particular technical qualifications are required, and as for the base stations above, the licensee may appoint others to do the on-air operation but he/she retains responsibility for the conduct of the station. Licences are issued by the Spectrum Management Agency.

If you wish to operate an outpost mobile, you are strongly recommended to register your call sign (from the licence) with the bases you intend to work through. The major operators of outpost radio bases are:

The Royal Flying Doctor Service, with bases at:

Alice Springs	Phone: (08) 8952 1033
Broken Hill	Phone: (08) 8088 0777
Cairns	Phone: (070) 531 952
Carnarvon	Phone: (08) 9941 1758
Charleville	Phone: (076) 54 1233
Derby	Phone: (08) 9191 1211
Kalgoorlie	Phone: (08) 9021 2211
Meekathara	Phone: (08) 9981 1107
Mt Isa	Phone: (077) 43 2800
Pt Augusta	Phone: (08) 8642 2044
Pt Hedland	Phone: (08) 9173 1386

Telstra Mobile Satellite and Radio Services, with bases at:

Sydney and Perth (national 24-hour operation), Melbourne, Brisbane, Townsville, Darwin. (Regional bases may have restricted hours.)

Direct all enquiries to:
Customer Service Centre
PO Box 153
CABOOLTURE Qld 4510
Phone: 1800 810 023

St John Ambulance Service of Northern Territory.
Base at Darwin:

Outpost Radio VJY
PO Box 40221
CASUARINA NT 0811
Phone: (08) 8922 6262

There are also several organisations which offer a service for boating and inland travellers with HF mobile transceivers (outpost radio type) based on telephone interconnection and general message handling. Individual services offerings range from a more or less private service based on a commercial company offering subscriptions, to a more communal group, for instance the members of a four-wheel drive explorers' club.

Broadcasting and television

In the broadcasting and television industry, there are many specialist careers in the fields of equipment operation and presentation of programs. The following are training organisations known to offer courses in these fields. (Note that these are mostly commercial organisations, so the normal provisions of *caveat emptor* apply. Inclusion of a name in this list is not intended as any sort of endorsement, recommendation or warranty of their offering, and lack of inclusion is not intended as any reflection on those who offer a similar service.)

Academy of Radio
PO Box 8319
PERTH BUSINESS CENTRE WA 6849
Phone: (08) 9227 9905

Announcer's Academy
120 Hawthorn Road
CAULFIELD Vic. 3162
Phone: (03) 9528 2777

Australian Film, Television and Radio School
Cnr Epping and Balaclava Roads
NORTH RYDE NSW 2113
Phone: (02) 9805 6611

Creative World Pty Ltd
51 Walker Street
NORTH SYDNEY NSW 2060
Phone: (02) 9959 4993

Graeme Lyndon Productions
11 Barwon Avenue
TURRAMURRA NSW 2074
Phone: (02) 9449 2591

Griffith University College of Art
Attn John Dommett
Film and Television Dept
Clearview Terrace
MORNINGSIDE Qld 4170
Phone: (07) 3836 3333
(only for video)

Max Rowley Media Academy
16 Terama Street
BILGOLA PLATEAU NSW 2107
Phone: (02) 9918 8000

Vaughan Harvey Radio School
33 Pirie Street
ADELAIDE SA 5000
Phone: (08) 8231 3087.

In addition to all of the above, all community radio stations have a training function for announcers, studio operators etc. These are listed in the Yellow Pages under the classification 'Radio stations'.

A.3 SOURCES OF INFORMATION FOR LESS FORMAL OR HOBBY PURPOSES

Amateur radio clubs

The major co-ordinating body for most amateur transmitting activity is the Wireless Institute of Australia (WIA) which has a Divisional Executive in each state capital. There is detailed information on the procedure for gaining an Amateur Operator's licence in the *Australian Radio Amateur Call Book*, published annually by the Wireless Institute.

Addresses for contact with WIA Divisional Offices are:

GPO Box 600
CANBERRA ACT 2601

PO Box 1066
PARRAMATTA NSW 2124
Phone: (02) 9689 2417

GPO Box 638
BRISBANE Qld 4001
Phone: (07) 3496 4714

GPO Box 1234
ADELAIDE SA 5001
Phone: (08) 8264 0463
(Includes both SA and NT)

5 Helm Street
NEWSTEAD Tas. 7250
Phone: (03) 6234 3553

40G Victory Boulevard
ASHBURTON Vic. 3147
Phone: (03) 9885 9261

PO Box 10
WEST PERTH WA 6872
Phone: (08) 9351 8873

There are also Web page addresses. For instance the address for the Adelaide office is:

http://www.vk5wia.ampr.org/

Citizen's band repeater operating groups

There are approximately 150 groups Australia-wide. These are mainly groups of non-technical people but the organising committees usually have contact with a source for technical information.

A comprehensive list of UHF CB repeater locations and licensing details can be obtained from the Spectrum Management Agency's 'RRL Database' in the form of a CD ROM copy. This listing records the licensee's name and postal address, but not telephone numbers. A copy of the disk can be obtained at a cost of $109 by ordering on a form obtainable from any office of the Spectrum Management Agency.

Other less comprehensive listings of repeater locations are obtainable from the following:

1. Scanner frequency listings, generally published as softcover books (one for each state) and sold through technical bookshops and radio hobbyist suppliers such as the 'Dick Smith' chain of stores.
2. Particular issues of CB specialist magazines. Contact the magazine publisher for information and back issues.
3. Each state office of the Spectrum Management Agency maintains a listing of repeater locations for the local state. This listing, however, shows only the location and gives no information on ownership or contact addresses.

School radio and electronics clubs

These clubs are mainly found in high schools; their locations depend more on the availability of an interested teacher or parent than on any geographic or demographic factors. In cases where the club operates a station in the amateur bands, the club is usually affiliated with the WIA (see above). There are no centralised listings of clubs associated with schools, but Head Teachers do talk to each other and the Head of a particular school will often be aware of the existence of specialist clubs such as this in other schools in the area.

Radio clubs associated with church or service club youth groups or large semigovernment or commercial organisations social clubs

If these clubs operate an amateur transmitting station, as with the school clubs they will usually be affiliated with the WIA; otherwise, there is no publicly available listing of them. In some local government areas, the council keeps a record of known clubs on an informal basis, or local libraries may be able to make an introductory contact for clubs in the local area.

Monitoring of international broadcasting stations

This activity is often engaged in by individuals who have little contact with a club or local group. It is particularly of interest to those who live in a radio-quiet location with room for directional aerials on the short-wave broadcasting bands (HF band).

Local amateur radio operators may be able to give you some useful information on monitoring and signal-reporting techniques. The WIA has some information on that aspect of the subject.

Most of the transmitters on the international bands are operated by national governments, so names and addresses can be found by referring to the Embassy, High Commission or Consul for that country.

There has also been a regular series of articles under the heading 'Shortwave Listening' running in the *Electronics Australia* magazine (one or two pages per issue) for many years. Reference to back issues of that magazine will give a wealth of incidental information about the mechanics of the hobby.

A.4 EMERGENCY SERVICES WHICH WELCOME VOLUNTEERS

Air Training Corps (ATC)

Initial contact should be through the RAAF Careers Office in the capital city of your state. ATCs are concerned with training in all aspects of aircraft operation, not just that part which deals with radio communication.

Army Reserve Signals Units

There is a recruiting centre in each state capital;
National contact: Phone 13 1902

These units deal with all aspects of telecommunications, and in some cases, radio operation may be only a minor part of their activity.

Australian Citizen's Radio Monitors (ACRM)

This is an organisation that is probably the least formalised of all the emergency services, but despite the informality, it often gives a particularly effective means of raising an alarm in the case of motor vehicle accidents on major roads. It relies on there being a large number of eyes and ears with access to citizen's band transceivers (both 27 MHz and UHF).

If you operate a CB radio and are interested in contacting ACRM, the best way may be to ask a few questions about the subject on air. It will not be long before your interest will be noticed by an active member of the local group.

There are also monitors accessible by telephone in each state capital city and in many of the larger country towns. Look for them in the alphabetical telephone directory under the name 'Australian Citizen's Radio Monitors'.

Coastguard

There is a continual demand for people who are able to join a flotilla and undertake training to become proficient in the whole range of seamanship and navigational skills required of a boat crew in adverse weather conditions and who can make time available at short notice for search and rescue duties. The training program for this proficiency usually takes about six months.

Radio operators are selected from those people who have a demonstrated aptitude for navigational and boat-handling skills; they are then given further training. All radio operators are required to hold the RROCP (see Section A.2). There are also opportunities from time to time for some people who are not fully trained to become associate members and to take on duties monitoring radio channels when 24-hour surveillance is required.

For initial contact with Coastguard groups, look first in the alphabetical listing of your local telephone directory. If you are seriously considering joining a group, the person to contact is the Squadron Commodore. If that approach is not fruitful, contact:

The National Commodore
Australian Volunteer Coast Guard Association
7 Newport Crescent
PORT MACQUARIE NSW 2444
Phone: (065) 83 1415

Country fire brigades

There is a fire station in almost every country town, usually sited with good access to several major roads, but most are only intermittently attended. There will often be a contact telephone number painted on the door of the brigade building—be careful though because the number most prominently displayed will usually be the fire call number and you may set off a very loud siren by ringing it! Brigades usually run a training session one night each week.

Naval Reserve Cadets

These have the same relationship to ships as Air Training Corps (see above) have to aircraft. Cadets undertake training in all aspects of ocean-related skills. There is a recruiting centre in each state capital.
National contact: Phone 13 1902.

Volunteer marine rescue groups

There are a number of local groups or state-based associations in seaside areas which, although they may be largely autonomous, share a common purpose of rapid response to emergency situations in coastal waters. All these groups work in close liaison with local police officers and State or Territory Emergency Service units, but in most cases the initial call for help comes via a radio channel. These groups have a need for both radio operators (who handle the emergency traffic during an operation) and radio monitors (who listen to the channel in case of a call).

There is an organisation named the Volunteer Marine Rescue Association of Australia, which acts as a forum for liaison between all the groups, who have contact with 235 groups (including the Coastguard flotillas) and can be used for initial contact with the nearest group to any location. That association is state-based and does not have a central head office for all of Australia.

As a back-up to that point of contact, the Australian Maritime Safety Authority (see Section A.2) maintains a list of all known organisations with sea rescue capabilities.

St John Ambulance Service

The particular expertise of the St John Ambulance Service is in providing effective first aid in large crowds such as at sporting events, parades or celebrations. Volunteers carry a belt-mounted portable transceiver. Great skill in radio operation is not usually required but volunteers must be prepared to work calmly in stressful situations with lots of bystanders looking on.

To contact this organisation, check first for listing in the alphabetical section of local telephone directories; otherwise, try:

National Headquarters
PO Box 3895
MANUKA ACT 2603
Phone: (06) 295 3777

State or Territory Emergency Services

The particular interest of State Emergency Service (SES) groups is those situations which are, or may become, of large enough scale to significantly disrupt the fabric of society in a particular area. SES groups work in close liaison with the Police Force and maintain contact with all other emergency-response groups. There is a state or territory headquarters in each of the capital cities, and initial contact can be made through telephone numbers listed in the alphabetical directory for the metropolitan area of each capital city. In country areas, the local police station will usually have information to direct you to the nearest active group.

GLOSSARY

A comparison between a listing here and a full dictionary definition would show that there are many words in this Glossary with more meanings than are listed here. The definition given here is that part of the full meaning which is particularly relevant to the electrical, electronic or radio branches of technology.

A

ASCII American Standard Code for Information Interchange

aerial an electrical conductor or related group of conductors arranged to efficiently couple energy between an electrical circuit and a radiation field; also may describe the supporting masts or towers for such a conductor arrangement

aerial gain increase in concentration of the power of a radiated or received signal which makes the beam of a directional aerial appear equivalent to an omnidirectional signal from a more powerful transmitter. A measurement in decibels for a particular aerial structure defines the amount by which transmitter power would have to be raised to give equal signal strength from an isotropic radiator

amateur operators people who use radio facilities and techniques for purposes other than commercial gain or paid work. Those who operate transmitters must hold the appropriate class of licence which, in Australia, is the AOCP or AOLCP, and must operate their equipment in bands designaed for amateur use

ampere (or amp) the unit of electrical current flow; indicates a particular number of electrons passing the measurement point each second

amplification literally 'making bigger'; in an electrical circuit it means the control of a power flow by a smaller signal (current or voltage) to produce an output which is a larger copy of the controlling signal

amplitude literally 'extension in space, especially breadth or width'; in an electrical circuit carrying an alternating signal, it is measured as the degree of current or voltage excursion of the signal

amplitude modulation a method of controlling the instantaneous power level of a radio carrier wave by a program or waveform of much lower frequency

analogue literally 'an agreement, likeness or correspondence'; used in electrical technology to describe a signal where amplitude, frequency response and timing of signal elements are all used to carry intelligent information. Analogue electrical circuits must be designed and operated so that the output is an undistorted copy of the input signal. Compare with 'digital'

antenna see also *aerial*; the meanings of these two words largely overlap but whereas 'aerial' always refers to the complete structure, 'antenna' may describe only the electrical conductors and may be used to concentrate the reader's attention on one particular conductive element

antenna gain see *aerial gain*

audible frequency a frequency in the range in which a sound wave can be heard by the human ear, usually reckoned for technical purposes to be between 16 hertz and 16 000 hertz

automatic gain control a circuit associated with an amplifier which senses the level or amplitude of the output and makes adjustments to an input control function to keep the output at a constant level

automatic link establishment a system for use, when multiple channels are available, which sends a test signal on each channel and selects for use the one which gives clearest transfer of information

B

back scatter literally 'the deflection of a wave by more than 90°'; usually used to describe a propagation mechanism which changes the direction of a minute proportion of the power of a signal by 180° so that a receiver placed at the point of transmission may detect a signal time-delayed by the distance travelled

balanced line a pair of electrical conductors arranged so that they carry equal and opposite currents and voltages so that radiation from either one is cancelled and prevented by an equal and opposite radiation flux from the other. Compare with *coaxial cable*

balanced modulator an electrical circuit in which two modulation processes are performed and the outputs compared. Those parts of the output which are the same in both are cancelled and only those parts which are different are passed on to the next stage. Most commonly used where there is a need to delete both the input signals from the output and pass on only modulation products

balun transformer an electrical component which is designed to have one side connected to a balanced line and the other to an unbalanced source or load and to efficiently transfer electrical power from one to the other

bandwidth the range of frequencies between the highest and lowest frequency of interest for, or relevant to, a particular function. See also *channel*, *passband*

baseband in a multiplexed transmission, the whole group of channels or signals which is modulated onto the radiofrequency carrier signal; the input signal to the final modulating stage in the transmitter or the output signal of the first demodulating stage of the receiver

beam in relation to directional aerials, the direction of the strongest transmitted signal or the most sensitive reception together with the small range of directions close to it where signal strength or sensitivity is approximately the same as for that most sensitive direction

beat frequency oscillator in a receiver designed for reception of single sideband or continuous wave transmissions, a local source of a steady oscillation which can be finely trimmed in frequency either to be exactly the same as that of the original suppressed carrier or to be a set amount different from it so that the carrier is revealed as an audible tone

bel the original unit of comparison of sound levels; numerically equal to 10 decibels

bessel functions a group of mathematical tools for calculating the frequency and relative power level of each of the sidebands of a frequency-modulated signal

bit error rate related to data transmission; the probability that any one of the most basic units of data will have its logic state reversed by any of the range of random processes that affect a transmission system

BMAC Type B multiplexed analogue component; a method of modulation specifically developed for transmission of television signals by geostationary satellites

broadband a relative term indicating width of radio spectrum space available for a particular purpose, usually indicating a transmission system with sufficient bandwidth for many channels of the type of information in question. The relevance of the term depends on the type of modulating signal; for instance a microwave link with an operating bandwidth of 5 MHz may be described as a broadband bearer for multiplexed telephony channels but not for a television signal because one channel of television signal will require almost all of the available channel space. However, a link with 20 MHz bandwidth could transmit several television signals simultaneously and so could be described as broadband for both telephony and television

broadcast transmission of information or a program to all who wish to receive it in a particular geographic area or region

C

call sign the 'name' of a radio station; a unique combination of letters and numbers which identifies transmissions. A call sign does not necessarily belong to a particular person or a particular transmitter or to a particular location, but may specify any of these at different times. If the station is mobile, the call sign usually specifies a particular vehicle, ship or aircraft

capacitance the property of a pair of conductive bodies which are held close to but insulated from each other to accept and store an electric charge. The amount of capacitance between two metal plates held parallel to each other depends on the distance between the plates, the overlapping area common to the two plates and a property called the 'dielectric constant' which relates to the material of the insulating layer. Capacitance can be roughly described as the electrical equivalent of mechanical elasticity of the insulating layer

capacitive reactance the ratio between applied voltage and resulting current when an alternating voltage is connected to a capacitive component; depends on the size of the capacitance and also on the frequency of alternation of the applied voltage

capture area the area of a sheet of perfectly absorbing material, placed perpendicular to the direction of propagation, which would absorb the same power from a radio wave as the aerial being measured

capture effect the ability of a receiver designed for FM reception to reduce the level of noise and interference in all cases where the wanted signal is stronger than all others

carrier frequency the frequency of alternation of a carrier wave

carrier wave an unvarying signal whose frequency is in the required place in the radio spectrum and which can be made to carry an intelligent message by modulation of either the instantaneous power level or the frequency

cat's whisker a component of one of the first successful forms of radiofrequency detector. The rectifying crystal required a minimum capacitance connection which was achieved by using the finest possible wire and a point contact

channel a portion of the radiofrequency spectrum just wide enough to accommodate the bandwidth of one signal. Actual width of a channel depends on the type of signal and partly on the method of modulation

characteristic impedance the ratio of voltage to current required in a transmission line to transmit power without reflection. The actual ratio is determined by materials and physical construction of the cable

circular polarisation a polarisation mechanism in which the magnetic and electric components of the wave rotate as the wave proceeds

clipper literally 'that which cuts'; an electrical circuit which refuses to respond to signals over a certain level; used in a transmitter as the last stage before the modulation process to prevent overmodulation

coaxial cable a radiofrequency transmission line in which one conductor is wrapped around the other with an insulating space between them. It is commonly used with the outer conductor at earth potential

coherent wave a wave motion in which there are no sudden phase changes. A sinusoidal wave is one example

collinear aerial an arrangement of aerial elements in which several are placed end to end and all are used as driven elements

comb filter an electrical circuit with a frequency response that has a number of passbands separated by stop bands so that a graph of the frequency response looks like the teeth of a comb. The passbands are often harmonically related

conductors substances or objects that allow easy passage of electric current. In an electrical circuit it can usually be assumed that all points connected by a conductor will be at the same voltage unless very heavy currents are flowing or the voltage is alternating with such a frequency that the size of the conductor is a significant proportion of the wavelength

convertor a circuit for changing frequency; the modulation of a signal on one channel is translated to a channel with a carrier wave of a different frequency. Also may be named 'mixer' or 'frequency changer'

coulomb the unit of quantity of electric charge; specifies a particular (very large) number of electrons

crossed dipoles two half-wave dipoles both used as driven elements placed perpendicular to each other so that the radiated signal has circular polarisation

crystal set originally meant a receiver that used a crystal of galena or similar, and a cat's whisker as the detector. In recent times the term is used to describe a receiver for AM broadcast reception which has only passive components and is entirely driven by power absorbed from the radiated wave

CW (continuous wave) signal a transmission in which the carrier wave is either transmitted at full power or switched off. Switching is usually under control of a hand key and intelligent modulation is transmitted by Morse or one of the related telegraphic codes

cycle one complete movement of an alternating signal, or movement from a starting point through the point of maximum excursion in each direction then returning to the starting point

D

data literally 'figures' or 'information'; in electrical terms, a flow of binary impulses which carries intelligent information to a computer or similar machine

day channel a channel in the HF band which gives reliable propagation during daylight hours between the two points required

decibel originally the unit of comparison of sound levels; now used for comparison of power levels in many different aspects of electronic technology. A logarithmic scale is used to match the sensitivity characteristic of human ears

de-emphasis using a frequency-response characteristic which slopes towards lower sensitivity or gain at the high-frequency end of the passband

demand assigned multiple access a method of sharing several radio channels between a large group of users such that no user has a particular allocation of any channel but a channel is allocated by a processor in a central control station at the time of use. Also may be called 'trunking' or 'dynamic channel allocation'

demodulation the general name for all processes by which the original modulated signal is revealed in a receiver. See also *detector*, *discriminator*, *product detection*, *ratio detector*

detection the form of demodulation most used for receiving amplitude-modulated signals

detector an electrical circuit usually based on the action of a rectifying diode for revealing the modulation of AM signals

deviation when used as a verb—changing the frequency of a carrier wave to perform the modulation process. The same term is also used as a noun or adjective to describe the measurement of frequency excursion resulting from the application of a particular modulating signal

dielectric an electrical insulator which allows electric forces to act through it; the insulating substance between the two plates of a capacitor

difference (or beat) frequency the frequency of that new component produced by an intermodulation process which can be calculated by taking the arithmetical difference of the frequencies of the two input signals

differentiation the mathematical function which has the effect that in time-related functions, the level of the output signal depends on the rate of change of the input

diffraction in optics, the mechanism that causes the shadows of sharply defined objects to have slightly blurred edges. In radio propagation, the mechanism that allows a signal to be received with reduced strength in the space that is geometrically shaded by an obstruction

digital literally 'resembling a finger'; a form of signalling in which only two valid states are recognised (finger is either held up or folded down) and information is transmitted by the timing of switching from one to the other. Circuits carrying digital information can be made relatively immune to amplitude-related noise and interference but require wider bandwidth to carry information at the same rate as an analogue signal

diode detector a detector for AM signals which uses the rectifying action of a diode

dipole aerial literally 'a pair of equal and opposite charges or forces'; an aerial element which has approximately equal lengths of conductor on each side of a central feed point

directivity the degree to which the output power or sensitivity of an aerial structure is concentrated into a small range of directions

director an aerial element which is placed in the intended direction of maximum sensitivity and cut to a length slightly shorter than the true resonant length which has the effect of increasing aerial gain in its direction

discriminator an electronic circuit whose output voltage depends in a linear fashion on the frequency of the input signal; used in the demodulator stage of receivers designed for frequency-modulated signals

distortion literally 'to twist out of shape'; the degree to which an output signal of an electronic circuit is not an exact copy of the input. The concept of twisting or misshaping applies to the waveshape of the voltage–time graph of the signal

Doppler effect an apparent change in wavelength or frequency observed when a transmitter and receiver are moving relative to each other

driven element a conductive element of an aerial which is electrically connected to the feeder either directly supplied with power from a transmitter or directly supplying signal to the input of a receiver

E

effective radiated power the radiofrequency power level that would be required to give the same field strength as that measured in the centre of the beam of a directional aerial if that aerial were replaced with an isotropic radiator

electrical length the length of a straight piece of conductor which would have the same electrical effect as the combination of conductors and lumped components under test; usually expressed as a proportion of the free space wavelength of the test signal

electromagnetic spectrum the total possible range of frequency–wavelength combinations for all radiated electromagnetic energy; includes all radio bands, all infrared, visible and ultraviolet light, and all the ionising radiations

electromotive force a physical force which can be detected by mechanical means resulting from the mutual attraction of separated positive and negative electric charges

electron the smallest and least massive subatomic particle that is definitely known to be capable of independent existence at rest; carries a unit negative electric charge

electron-volt a unit of quantity of energy equal to the energy change of one electron falling through a potential difference of one volt

element (of an aerial array) a single length of electrical conductor of the appropriate size and in the required position to have an effect on the directional properties or forward gain of an aerial structure

energy of a quantum (of radiation) the quantity of energy (usually measured in electron-volts) required to exactly change an orbiting electron from one permitted energy level to another. In the radiated wave, the energy of each quantum depends on the frequency of the wave, so only waves of a few particular frequencies can be absorbed by each electron

error correction techniques a range of technical operations used with transmission of digital signals which involve adding extra information to the bit stream before transmission so that the truth of blocks of data can be tested at the receiver

F

fading temporary reduction of strength of a received radio signal which may be due to any one of a wide variety of causes

feed lines transmission lines

field a region of space influenced by some force or energy

field strength for radiating signals, the concentration of power flux per unit of area perpendicular to the direction of propagation; can be expressed in units such as watts per square metre but also may be expressed as an equivalent voltage per linear measure

filter an electrical or electronic circuit using components with frequency-dependent properties in such a combination that the frequency/attenuation response graph shows a flat passband but rapidly changes to high attenuation for frequencies outside the passband

forward error correction an error-correction technique in which sufficient extra information is added to the transmitted signal to allow a corrupted signal to be checked and reconstituted at the receiver without further transmission of information

forward scatter a radio propagation mechanism in which a minute proportion of the transmitted signal has its direction of travel changed sufficiently to provide for reception at distances several times the distance to the visible horizon

free space a volume of space which contains no electrical conductors, no absorbers of radiated energy and no materials that will slow the speed or change the direction of the radiated wave

frequency literally 'rate of recurrence'; in electrical use, the number of cycles of an alternating wave completed in each unit of time (usually one second)

frequency modulation a process whereby the instantaneous voltage level of an intelligent program or signal (in electrical form) is used to slightly vary the frequency of a constant-amplitude radio transmission

fringe area the geographical region in which a transmission can be received but with marginal signal strength or reduced reliability

front to back ratio the power ratio (usually expressed in decibels) between the signal from an aerial at the centre of the main beam and the signal radiated in exactly the opposite direction

G

gain the ratio by which the output of an amplifier or similar electronic device exceeds the input; may be expressed as voltage gain, current gain or power gain; often measured in decibels. See also *aerial gain*

geostationary orbit one of a family of orbits for artificial Earth satellites for which the orbital inclination is zero degrees with respect to the equator, and the orbital period is exactly 24 hours. The practical effect is that the satellite appears to hang motionless above a particular spot on the Earth's surface and can be linked with high-gain aerials which do not need to be moved to track the satellite

ghost image on a television receiver, the fainter image—displaced usually to the right—that results from signals arriving by two or more different paths which have different lengths and therefore different transit times

ground wave a signal which is propagated by following the interface layer between the electrically insulating atmosphere and the partially conductive ground and therefore tends to follow a curved earth surface

group delay a function of differences in total transmission time for portions of a broadband at different frequencies; if incorrect, causes smearing of the edges of pulses

H

heterodynes the new frequencies that appear in the output of a mixer or detector when the input contains two frequency components which are able to intermodulate

horizontal polarisation the polarisation of a signal radiated from an aerial with the driven element placed parallel to the Earth's surface

hundred per cent modulation for amplitude-modulated signals, the special case where level of modulation is just sufficient to reduce carrier output to zero at the bottom of the trough; for other forms of modulation, the level of modulating signal which is just sufficient to use all of the allocated bandwidth or rated power

Huygens's principle the theorem that a wavefront propagates in a direction perpendicular to the plane of phase equality. This theorem satisfactorily explains the change of direction due to change of propagation velocity that is observed in the refraction process

hybrid an electrical component that has connections for a two-wire line (for instance a telephone line) on one side and provision for a four-wire connection on the other, with high isolation between the send and receive pairs of the four-wire connection

I

IF amplifier all the components of a receiver which form a fixed tuned high-gain amplifier between the output of the mixer and the input of the demodulator. It provides most of the amplification and adjacent channel selectivity of the whole receiver

image frequencies the frequencies of unwanted channels of sensitivity of a receiver resulting from the fact that mixing processes are equally sensitive to both upper and lower sidebands

impedance the ratio of voltage to current without reference to phase; compare with *reactance* and *resistance*

inductance the energy storage factor that operates when an electric current creates a magnetic field; in outline, may be described as the electrical equivalent of mechanical inertia

induction field the region of space close to a radiating conductor where inductive and capacitive fields are predominant, and energy is being swapped back and forth between the conductor and the field

inductive reactance the ratio of voltage to current which is observed when an alternating voltage is applied to a component which stores energy in a magnetic field. Its value depends on the inductance of the magnetic component and also on the frequency of alternation of the applied voltage

insulators materials which allow very little current flow when an electrical voltage is applied across them; the word may also describe particular-shaped items of such material which are designed to hold conductors in position without leakage of current

intermediate frequency a fixed channel or band usually higher in frequency than the output of the detector or demodulator stage and lower than the radiofrequency, which gives opportunity for trimming of filter characteristics for flat passband response and high adjacent channel selectivity

inverse square law the principle (originally from the field of optics) that in free space the intensity of a signal diminishes with increasing distance in accordance with the square of the distance from the source

ionisation the process of adding sufficient energy to an orbiting electron so that it is removed from its orbit, or the process of adding an electron at the right energy level so that it is attached to an atom. In either case, the whole atom behaves as an electrically charged particle

isotope an atom of a chemical element which has other than the usual number of neutrons; see also *radioisotopes*

isotropic radiator a mathematically fictional device which gives exactly even radiation in all directions so that a hollow sphere with its centre at the central point of the radiator would have even illumination over the whole of its inner surface

J

jitter short-term variations of timing or phase of a data signal that reduce the definition of the transitions of logic state

joule the SI unit of quantity of energy equivalent to a power flow of one watt for one second of time

K

Kelvin temperature a temperature measurement scale whose zero reference is that temperature at which all thermal movement of atoms has ceased

L

limiter

1. in a broadcast transmitting station, the piece of equipment which senses the instantaneous level of the program and makes short-term adjustments to total system gain to prevent overmodulation
2. in a receiver designed for FM reception, an amplifier (usually part of the IF amplifier) which refuses to give output over a certain level no matter how high the level of the input; this has the effect of removing amplitude-related interference from the output signal

link literally 'to join up or establish communication'; a two-way channel of communication between two particular points formed by a transmitter and receiver at each end with the receivers tuned to each other's transmissions

load (termination) a resistive component connected to the end of a transmission line of the correct value to absorb all of the power as it arrives without any being reflected. A load may also be an aerial which is tuned to a purely resistive input impedance

loading coil an inductive component added to the base of a whip aerial to adjust the electrical length for correct matching to a transmission line or transmitter output

local oscillator a source of radiofrequency signal built into a superheterodyne receiver which can be adjusted in frequency so that the output of the associated mixer stage is in the passband of the following IF amplifier

logic states in a digital transmission system, either of the two voltage levels which have significance to the device receiving the binary information

lower sideband the band of frequencies between the carrier frequency and the low-frequency limit of the channel in which modulation products that are lower in frequency than the carrier wave are transmitted

M

magnetic field the region of space close to a magnet, or to a conductor carrying an electric current, in which energy is stored and a physical force can be detected by its effect on another magnet

matching (of transmission lines) the adjustment of a load or tuning of a transmitter to achieve maximum power transfer without reflection

maximum usable frequency the radiofrequency in the HF band for which those frequencies slightly lower can be received by sky-wave propagation with strong clear signals, and for which all those above it cannot be received at all. The actual frequency is highly variable and depends on the locations of transmitter and receiver, time of day, season of year and the degree of recent activity on the Sun's surface

microwave literally 'waves of microscopic size'; originally used to describe all those channels in the spectrum where the wavelength was so short that only signals over line-of-sight paths were to be expected. In more recent times, a more restricted use of the word has become common; it is now used to describe those channels for which waveguide is normally used as a transmission line

mixer frequency changer or convertor; in single sideband transceivers, a mixer circuit may also be used as the modulator in the transmit direction

modem an abbreviation of 'modulator/demodulator'; an electronic device which can simultaneously handle signals in both directions, taking signals in the send direction and modulating them onto a carrier or subcarrier and at the same time receiving a modulated signal and producing a demodulated output

modulate literally 'to regulate or adjust'; to control a property of a carrier wave (either amplitude or frequency) in step with the instantaneous level of an intelligent signal

modulating frequencies the range of frequencies, such as an audio program or composite video or data baseband, which specifies the input channel bandwidth of a modulation process

morse code the first successful method of transmitting alphabetic and punctuation characters achieved by simply switching the transmission medium on or off. It uses a coded sequence of short and long pulses to specify each character and may be used with any transmission medium

multiplexing sending many messages or programs simultaneously over the same piece of wire or communication channel

N

night channel a channel in the high-frequency band which gives reliable propagation between the two points required during the hours of night-time

noise a combination of sounds or signals from many unrelated sources which contains no intelligence of its own and places an ultimate limit on the usable sensitivity of a communication channel or listening device

noise limiting any of a number of electronic techniques to isolate a feature of an intelligent message and to use that feature to reduce the effect of non-intelligent components of the received signal

notch filter an electronic circuit which has high attenuation for frequencies in a very narrow range and passes all others with little change

O

Ohm's law the principle that current through an electrical component depends on the voltage applied and on an intrinsic property of resistance of the component; it is the most basic relationship of electrical theory

oscillation literally 'a steady swing—simple harmonic motion'; in electronics, a continuous train of coherent waves of alternation of voltage or current; the output of an oscillator circuit

oscillator an electronic circuit for producing a signal which is a steady flow of cycles (constituting a coherent wave) of alternating voltage or current

overmodulation overloading of the modulation performance of a transmitter so that sidebands or extra components of the transmitted signal are produced thereby causing adjacent channel interference

P

parallel tuned circuit a combination of electronic components, chiefly inductance and capacitance, in which the components are placed in parallel and the cancellation of reactances at frequencies close to resonance serves to increase the overall impedance of the combination to a very large value

parasitic elements aerial elements which are not electrically connected to the transmission line but which affect the directional properties of the aerial by absorbing and reradiating power from the driven element(s)

parity bit an extra logic element added to each data word before it is transmitted which is adjusted in logic state so that the combined states of all bits in that word are always either odd or even

passband the range of frequencies within which a filter or tuned amplifier will transmit signals with little attenuation or its intended amount of gain

passive components those components of an electrical circuit such as resistors, capacitors and inductances which have no amplifying or rectifying properties

peak voltage the highest potential difference reached at any instant of the cycle of alternation; important for the calculation of insulation requirements usually so reckoned as the potential difference between two nominated conductive points

phase modulation a form of transmission in which small changes to the absolute timing of cycles of the waveform are controlled by the modulating signal. Phase-modulated signals can be related in their characteristics to frequency-modulated signals

phonetic alphabet a list of words used to unambiguously specify letters of the alphabet; used either when the communication channel is affected by noise or when the spelling of words is a significant part of the message

pixel the smallest unit of information of a visual display; a signal element on one line of a raster which, if displayed in a contrasting colour, is seen by an observer as one indefinitely small dot

polar diagram a graphical presentation of the directional properties of an aerial in which direction from a central point is related to either compass bearing or direction in elevation, and signal strength is defined by relative distance from the central point

polarisation literally 'confining vibrations to a single plane'; in relation to radio aerials, the factor that causes a signal launched by, for instance, a horizontal driven element to be severely attenuated by an attempt to receive it with an aerial aligned vertically

potential difference the electrical pressure (voltage) between two nominated points

pre-emphasis using a frequency-response characteristic which slopes towards higher sensitivity or gain at the high-frequency end of the passband. Most commonly used in FM broadcasting transmitters to maximise signal-to-noise ratio of the treble part of a program

preselection tuned circuits used in a superheterodyne receiver before the frequency changer stage whose purpose is to reduce sensitivity to image frequencies

primary service area the geographical region close to a broadcasting transmitter station where it is expected that noise-free reception should be possible in all locations

product detection an electrical function which is equivalent to the mathematical operation of multiplying two signals together. The electrical process is similar to amplitude modulation and produces sidebands which can be arranged so that for one of the sidebands the carrier is exactly translated to zero frequency

propagation literally 'to cause to extend to a greater distance'; for radio signals, it is the general term that describes all the processes and mechanisms by which signals are carried from the transmit to the receive aerials

proton a subatomic particle which resides in the nucleus of an atom and carries an electrical charge equal to that of an electron but opposite in effect. Each chemical element has a particular number of protons in the nuclei of its atoms, and it is this that determines the chemical properties of the element

pro-words single spoken words for which all users of a particular radio service have agreed that when they are used other than in a sentence, they will indicate a particular function of the radio service; for instance, 'over' used on a two-way simplex service indicates 'reverse the direction of transmission'

pulse code modulation a generic term to describe forms of transmission of pulsed radio signals in which a characteristic of the pulse is controlled by the modulating signal. A wide variety of coding algorithms are possible

pulse transmission a form of radio transmission in which all the power is transmitted in very short bursts of high amplitude with relatively long times between pulses during which no power is radiated at all. Its advantage is that the heating effect on transmitter components is related to average power flow but with a properly designed and adjusted receiver, signal-to-noise ratio can be determined by the peak level of the pulses. However, it requires a wide bandwidth to accommodate the very short pulses

Q

quantum the amount of energy required to exactly move an orbiting electron from one permitted energy level to another. Also describes an electromagnetic wave having the form of a damped train of oscillation which contains exactly that amount of energy

quarter-wave matching section a section of transmission line which is electrically a quarter-wavelength and whose characteristic impedance is the geometric mean between a source of radiofrequency power and a load. (The source and load in this case could be other sections of transmission line)

R

radar short for '*ra*dio *d*etecting *a*nd *r*anging'; a method of locating the geographical position of objects or vessels by sensing radiated energy reflected from them

radiant energy a flow of energy outwards from a source which continues independently of the source and would continue even if the radiation process ceased and the source were removed

radiation literally 'emission outwards from a source'; for electromagnetic waves, refers to those which have gained independence from the originating electrical circuit or source and continue to transfer energy with no physical connection to the source

radio the technological process of using radiated electromagnetic energy to carry intelligent messages or programs

radio spectrum that part of the electromagnetic spectrum of wavelengths longer than those in the infrared region in which coherent waves can be generated and modulated to carry intelligent messages or programs

radioisotopes an isotope in which the number of neutrons is such that the atom is unstable and will at some time in the future either emit one or more of the excess neutrons or split into two atoms of other chemical elements. The splitting or emission process is described as 'radioactive decay' and is accompanied by radiation or absorption of large amounts of energy in the form of ionising radiation

radiolocation any one of a number of techniques for fixing the geographical position of an object or location and which uses the physical characteristics of radio waves for measurement

radionavigation an aspect of the techniques of radiolocation which is directed to the positioning and guidance of ships and aircraft

radiophysics a branch of the physical sciences which deals with the properties and propagation of radio waves

radiotelegraphy a form of message transmission in which the radio wave is switched on and off in accordance with a code. CW (continuous wave) transmission of characters by Morse code is one example

ratio detector an electronic circuit which is sensitive to variations of frequency of an incoming signal compared with the frequency of exact tuning of a reference tuned circuit. Used for demodulation of FM signals and gives some rejection of the effect of amplitude variations

reactance impedance in a capacitive or inductive component; the ratio of voltage to current in all cases where there is a phase shift between them of 90°

receiver an electronic instrument which is able to select a particular signal and detect the modulation of it from the range of signals appearing at the output of an aerial placed in a radiation field. Most receivers also provide for amplification of the wanted signal and some other signal-processing functions

recharge effect the ability of a large capacitor to store a charge in a way that cannot be discharged quickly, so that if the normal connection terminals are discharged but then left open circuit, a possibly lethal charge will reappear on the open-circuited terminals

reciprocity theorem a principle which applies to all passive transducers, but in the radio field, is particularly relevant to aerials; energy flow through the transducer is bidirectional so that all performance measurements (such as gain or directivity) derived from a transmitting aerial can be equally applied to a receiving aerial, and vice versa

reduced carrier a method of modulation similar to single sideband in which a proportion of the power of the carrier is transmitted along with the sideband to avoid the need for a critically tuned signal to replace the carrier in the receiver

reflection

1. the change of direction of a light beam or radio signal resulting from contact with a medium of changed refractive index. The laws of optical reflection apply equally to all wave motions
2. the reversal of direction of a portion of the power flow in a transmission line resulting from a mismatch between the impedance of the load and the characteristic impedance of the line

refraction a change of direction of propagation of a plane wavefront when a part of the plane advances with an altered velocity of propagation. Radio waves obey the laws of optical refraction but with some changes of emphasis due to the vastly different wavelength

repeater a combination of transmitter and receiver arranged so that the output of the receiver is connected to the input to the transmitter and retransmitted. Usually located on a prominent position so that the area or distance covered is increased by the retransmission

resistance impedance in which voltage and current are in phase and electrical power is absorbed and converted into heat energy; the ratio of voltage to current in a component in which power is being absorbed

resonance the appearance of a relatively large amplitude vibration or oscillation by the storage of energy which occurs when a system with a natural frequency (such as a tuned circuit) is excited by an alternating signal with frequency close to that natural frequency

return loss the ratio of power levels (usually measured in decibels) in the incident and reflected signals in a transmission line

S

satellite (artificial) artificial objects which have been placed in fixed orbits around the Earth. Communication with them is via radio link and some may be used to provide a radio repeater function to give very long-range capabilities for some channels in the UHF and microwave bands

scanning receivers radio receivers with multichannel capability which can be rapidly switched from channel to channel by an automatic switch, with the switching process being suspended whenever an occupied channel is found

scattering dispersion of the direction of propagation of a minute part of the power of radio signals due to dust particles and air mass boundaries. See also *back scatter* or *forward scatter*

selectivity the degree to which a tuned circuit or filter can enhance the power level of signals within the passband compared with that of signals of other frequencies

semiconductors

1. materials which have electrical conductivity intermediate between that of conductors and insulators and in which the conductivity can be made to depend on the direction of applied voltage by introducing traces of other elements into the crystal lattice
2. electrical and electronic components made from semiconductor materials which have functions related to rectification or amplification of signals

series tuned circuit a combination of electronic components, chiefly inductance and capacitance, in which the components are placed in series and the cancellation of reactances at frequencies close to resonance serves to reduce the overall impedance of the combination to a very small value

shape factor the ratio between the bandwidth of a tuned filter at 60 decibels rejection compared with the bandwidth at 6 decibels rejection. Gives information on the ability of the filter to provide a flat passband for the wanted signal while at the same time giving good rejection of closely adjacent channels

sideband a band of frequencies close to the carrier frequency of a modulated signal in which the products of modulation are transmitted

signal literally 'an intelligent sign or message of any sort'; in radio use, a transmission which carries or could carry an intelligent message or program

signal-to-noise ratio the number of times by which a wanted signal is stronger than the total of all other energy inputs received on that channel

sine waves a wave motion in which the instantaneous position of a particle of the wave can be specified by the relationship of an angle to its sine. In practice, it is the only waveform which can be described by a single frequency; all other waveforms can be analysed into a combination of sine waves with a range of different frequencies

single sideband (SSB) a form of modulation derived from amplitude modulation in which only one of the two sidebands is transmitted; the carrier and the other sideband are prevented from reaching the transmitter output

single sideband transceiver an item of electronic equipment commonly used on the HF band for point-to-point or point-to-mobile communication which uses the single sideband method of modulation and in which a significant part of the circuit is used for both transmitting and receiving

skip signals see *sky wave*

skip zone the geographical region close to a transmitter operating in the HF band which is too close for communication on the frequency being transmitted because the signal is skipping over the receiving location

sky wave a method of propagation of most significance to the MF and HF bands in which signals are returned to Earth at long distances after reflection from a layer of electrically charged particles in the upper atmosphere

slotted line an instrument which contains a section of coaxial transmission line with a slot in the outer conductor through which a voltage-detecting probe can be inserted; used in the laboratory for directly measuring and displaying standing-wave ratio

spark transmissions radio signals generated by applying electrical power to a spark gap associated with a tuned circuit

spectrum the literal meaning of this word is derived from '**spectre**', meaning 'a visible apparition'. Originally described the apparition of colours from white light passed through a prism; now used to describe the group characteristics of a broad range of frequencies or wavelengths. See also *electromagnetic spectrum* and *radio spectrum*

spectrum analysis the action of displaying in graphical form the comparative power levels of signals on a range of frequencies; usually displayed with frequency plotted along the x-axis and power level on the y-axis

speed of light the velocity of propagation of electromagnetic waves which, in a vacuum, is close to 299 700 kilometres per second but slowed in dense mediums by an amount depending on the density of the medium and the frequency of the wave motion

splatter the effect of extra sidebands outside the allocated channel of a transmission produced by a distortion or overmodulation process

standing wave the wave-like shape of a graphical presentation of voltage or current measurements taken at a range of distances along a transmission line which is supplying power to a mismatched load

standing-wave ratio the ratio between largest and smallest voltages or currents that can be measured by testing at all points on a transmission line which is supplying power to a mismatched load

stub matching a method of matching a load to a transmission line in which a section of line close to, but not exactly, a quarter-wavelength is connected in parallel with the load and trimmed in length to provide the effect of a reactive component which cancels the effect of reactance in the load

subcarrier a component of a multiplexed signal which has the effect of providing a carrier frequency reference for one of the signals being transmitted

sum frequency the frequency of that new component produced by an intermodulation process which can be calculated by taking the arithmetical sum of the frequencies of the two input signals

superheterodyne receiver a receiver which uses a frequency changer or mixer stage to translate input signals on a range of frequencies to a single intermediate frequency for filtering and amplification

synchronous detection an electronic process similar to product detection in which the signal to be demodulated is compared to a locally generated reference signal and only those components which constitute the original sidebands are passed on as output

T

talk power describes the ability of a limiter or clipper to raise the average level of modulation without causing adjacent channel interference

telegraphy transmission of text, digits and punctuation characters by coded pulses (of, for instance, carrier power switched by a Morse key)

telemetry literally 'taking measurements over a long distance'; any of a wide range of systems or processes for transferring measurements from an unattended measuring device to a monitoring station by use of electrical signals on a wire circuit, or data signals on a radio link

the skip colloquial name for signals propagated by sky wave

threshold level the minimum signal level required at the input of an FM receiver to overcome receiver internal noise and to ensure that the capture effect works in favour of the external signal

time division multiplexing a method of transmitting several or many data signals on a single broadband bearer which uses synchronised switching of channels at transmitter and receiver, and speed buffering of each data channel

tracking adjustment of tuning rate related to mechanical movement of a control so that local oscillator and preselector tuned circuits of a tunable superheterodyne receiver always select the same input signal

transmission lines electrical conductors arranged in a variety of configurations for transfer of radiofrequency energy from place to place with minimum radiation. See also *balanced line*, *coaxial cable* and *waveguide*

tropospheric scatter links broadband bearer systems carrying telephony and/or data which use UHF or SHF signals in an over-the-horizon mode by scattering in the upper part of the turbulent weather zone of the atmosphere

tuned amplifier an electronic circuit which combines both frequency-dependent components and an amplifier so that a small proportion of the spectrum of input signals is amplified and all other frequencies are either passed at zero gain or reduced in level at the output

tuned circuit an electrical circuit, containing capacitive and inductive components, which exhibits electrical resonance and gives a very much increased response to a small band of frequencies close to the resonant frequency

U

upper side band the band of frequencies between the carrier frequency and the high-frequency limit of the channel in which modulation products that are higher in frequency than the carrier wave are transmitted

V

VF (voice frequency) channel the range of frequencies between approximately 300 and 3000 hertz which is just sufficient to transmit a message by human voice with maximum intelligibility in cases where the sensitivity of the channel is limited by random noise broadly scattered over the spectrum

volt the unit of electromotive force or electrical pressure; originally defined as the open circuit electromotive force of a cell made with a particular combination of chemicals; now defined in the SI system of measurements by the heating effect of a current flow

W

watt the unit of power or energy flow or the rate of doing work; in the SI system of measurements, defined as the power flow of one joule per second

waveform the shape of one cycle of a wave such as would be displayed when the wave is applied to the input of a properly adjusted oscilloscope

wavefront a figure, which is a plane in the case of a three-dimensional wave or a line in the case of a surface wave, which travels with the wave and connects together all points of equal phase

waveguide a piece of electrically conductive tubing of the correct size (approximately half the wavelength in its widest dimension) to be used as a transmission line with little loss of signal

wavelength the length of one complete cycle of a wave; the distance, measured in the direction of propagation, between a reference point on one wave and the next point that has exactly the same relative phase

whistlers (VLF) signals of such a frequency that they can be heard as audio tones without demodulation. They are received by using very long wire aerials and sensitive audio amplifiers. Their origin is not known and no practical use is made of them at this time

wireless originally 'wireless telegraph'. Electronic equipment and systems for transferring messages from one location to another with no physical connection between them by using radio waves

wireless LAN a form of local area network for computer interconnection in which communication to the terminal is achieved with a short-range broadband radio link operating in the super-high frequency (SHF) or extremely high frequency (EHF) bands

Y

Yagi a form of directional aerial which uses one driven element, usually one parasitic reflector, and from one to about eight to ten parasitic directors to achieve pronounced directional effects and significant forward gain

Z

zero decibels the reference for power level comparison which may be the level of the input to an amplifier which the output is referred to for gain measurement, or the level of a reference signal before a particular operation such as tuning adjustment is performed

INDEX

M

P

S

T

U

V